AF363305

Information Theory in Molecular Evolution

Information Theory in Molecular Evolution: From Models to Structures and Dynamics

Editor

Faruck Morcos

MDPI • Basel • Beijing • Wuhan • Barcelona • Belgrade • Manchester • Tokyo • Cluj • Tianjin

Editor
Faruck Morcos
Biological Sciences
University of Texas at Dallas
Richardson
United States

Editorial Office
MDPI
St. Alban-Anlage 66
4052 Basel, Switzerland

This is a reprint of articles from the Special Issue published online in the open access journal *Entropy* (ISSN 1099-4300) (available at: www.mdpi.com/journal/entropy/special_issues/inf_theory_Mol_Evol).

For citation purposes, cite each article independently as indicated on the article page online and as indicated below:

LastName, A.A.; LastName, B.B.; LastName, C.C. Article Title. *Journal Name* **Year**, *Volume Number*, Page Range.

ISBN 978-3-0365-1213-6 (Hbk)
ISBN 978-3-0365-1212-9 (PDF)

Contents

About the Editor . . . . **vii**

Preface to "Information Theory in Molecular Evolution: From Models to Structures and Dynamics" . . . . **ix**

Faruck Morcos
Information Theory in Molecular Evolution: From Models to Structures and Dynamics
Reprinted from: *Entropy* **2021**, *23*, 482, doi:10.3390/e23040482 . . . . **1**

Adam J. Hockenberry and Claus O. Wilke
Phylogenetic Weighting Does Little to Improve the Accuracy of Evolutionary Coupling Analyses
Reprinted from: *Entropy* **2019**, *21*, 1000, doi:10.3390/e21101000 . . . . **5**

Edwin Rodriguez Horta, Pierre Barrat-Charlaix and Martin Weigt
Toward Inferring Potts Models for Phylogenetically Correlated Sequence Data
Reprinted from: *Entropy* **2019**, *21*, 1090, doi:10.3390/e21111090 . . . . **21**

Duccio Malinverni and Alessandro Barducci
Coevolutionary Analysis of Protein Subfamilies by Sequence Reweighting
Reprinted from: *Entropy* **2019**, *21*, 1127, doi:10.3390/e21111127 . . . . **41**

Paul Campitelli and S. Banu Ozkan
Allostery and Epistasis: Emergent Properties of Anisotropic Networks
Reprinted from: *Entropy* **2020**, *22*, 667, doi:10.3390/e22060667 . . . . **55**

Feng Wang, Hongyu Zhou, Xinlei Wang and Peng Tao
Dynamical Behavior of -Lactamases and Penicillin- Binding Proteins in Different Functional States and Its Potential Role in Evolution
Reprinted from: *Entropy* **2019**, *21*, 1130, doi:10.3390/e21111130 . . . . **73**

Xavier F. Cadet, Reda Dehak, Sang Peter Chin and Miloud Bessafi
Non-Linear Dynamics Analysis of Protein Sequences. *Application to CYP450*
Reprinted from: *Entropy* **2019**, *21*, 852, doi:10.3390/e21090852 . . . . **99**

Claude Sinner, Cheyenne Ziegler, Yun Ho Jung, Xianli Jiang and Faruck Morcos
ELIHKSIR Web Server: Evolutionary Links Inferred for Histidine Kinase Sensors Interacting with Response Regulators
Reprinted from: *Entropy* **2021**, *23*, 170, doi:10.3390/e23020170 . . . . **119**

About the Editor

Faruck Morcos

Dr. Faruck Morcos is an Assistant Professor in the Departments of Biological Sciences, Bioengineering, and the Center for Systems Biology at the University of Texas at Dallas. He joined UTD after completing his postdoctoral training at the University of California San Diego, and Rice University. He is the recipient of multiple awards including the Research Excellence Award given by the CSE Department at the University of Notre Dame for his PhD thesis, and the Werner von Siemens Excellence Award for graduate studies at the Technical University of Munich. Dr. Morcos directs the Evolutionary Information Lab which focuses on solving problems at the interface between biology, computation, information theory, and biological physics. He received the NSF CAREER Award and the NIH MIRA Award to develop methods to extract biological information from sequence and genomic data and create models for molecular evolution, protein structure, function, and design, as well as bimolecular interactions.

Preface to "Information Theory in Molecular Evolution: From Models to Structures and Dynamics"

Modern biological sciences are driven by information. Large amounts of experimental data are collected and synthesized to create models to explain the complexity of biological systems. In addition, inter- and intra-cellular information processing is key to understanding cellular physiology and disease. The study of evolution and, in particular, molecular evolution, has benefited from information theoretical insights since the foundational work of Ronald A. Fisher. In recent years, there has been a growing interest in using tools from information theory and statistical physics to quantify and model the evolutionary processes. An integration of quantitative evolutionary models with structural aspects of biomolecules has energized scientific contributions and discovery. Applications include: the fields of protein structure prediction; protein folding; conformational plasticity in molecules; chromosome architecture and epistasis. Modern approaches also look at the study of dynamics, allostery, and interactions within complexes that facilitate molecular recognition and catalytic specificity.

This issue includes contributions from scientists with diverse and interdisciplinary backgrounds and aims to be accessible to a wide range of scientists including graduate students, postdoctoral researchers and principal investigators interested in quantitative aspects of molecular evolution driven by the analysis or large amounts of data and models.

Faruck Morcos
Editor

Editorial

Information Theory in Molecular Evolution: From Models to Structures and Dynamics

Faruck Morcos [1,2,3]

1 Department of Biological Sciences, University of Texas at Dallas, Richardson, TX 75080, USA; faruckm@utdallas.edu
2 Department of Bioengineering, University of Texas at Dallas, Richardson, TX 75080, USA
3 Center for Systems Biology, University of Texas at Dallas, Richardson, TX 75080, USA

Citation: Morcos, F. Information Theory in Molecular Evolution: From Models to Structures and Dynamics. *Entropy* **2021**, *23*, 482. https:// doi.org/10.3390/e23040482

Received: 15 April 2021
Accepted: 15 April 2021
Published: 19 April 2021

Publisher's Note: MDPI stays neutral with regard to jurisdictional claims in published maps and institutional affiliations.

Historically, information theory has been closely interconnected with evolutionary theory. The work of Ronald Fisher in population genetics [1] and the formulation of the principle of minimum Fisher information [2] are just two early examples of such connections. In recent years, with the advent of high-throughput sequencing technologies, the field of molecular evolution has been able to take advantage of large amounts of samples from evolution to improve models and applications to understand structural, dynamical, and functional aspects of biomolecules. Information metrics have been prevalent, in recent years, to estimate the likelihood that two amino acid sites in a protein are coevolving. A relevant example of such metrics is *Direct Information* (DI) [3,4] used in the context of Direct Coupling Analysis to estimate if two positions in a multiple sequence alignment are likely to be proximal in the 3D structure of a protein or RNA molecule. Other standard information metrics like *Mutual Information* have been applied and are particularly useful for the case of molecular complexes and interactions [5,6].

This special issue focuses on important aspects of the study of molecular evolution through the statistical features of sequence data, molecular simulation, and evolutionary convergence towards specificity in signaling networks. Three articles [7–9] investigate how phylogenetic relationships in sequence data have an effect in the inference procedure of a joint probability distribution $P(a_1, a_2, a_3, \dots, a_L)$ of a given sequence of length L. Particularly, these studies are centered under the premise that a preprocessing step for multiple sequence alignment analysis might reduce phylogenetic bias and could improve the inference procedure. These methods, ultimately, improve prediction of amino acid contacts and functional connections among amino acid sites.

In [7], Hockenberry et al. conducted a systematic study of previously relevant reweighting schemes that have been useful in other applications. These methods contrast versus the current practices of identity-based sequence reweighting used in Potts model inference. They find that previous applications do not add considerable value for the inference task and leave open the question for novel schemes that might improve the inference of coevolving residue pairs. Interestingly, in [8,9], the authors propose novel schemes to account for phylogenetic bias. First, Horta et al. [8] introduce a new inference method which uses a priori information about phylogeny to enhance contact prediction and fitness effects in simulated data. Second, Maliverni et al. [9] propose another scheme called continuous sequence reweighting (SR) that reveals structural features that are unique to subfamilies as opposed to determining global properties common to all family members. These articles as a whole provide an in-depth and useful picture on how to deal with phylogenetic correlations in the task of contact inference and the estimation of the effects of mutation.

A second set of articles in this issue [10–12] deals with the complex problem of evolutionary dynamics in protein structures and sequences. Cadet et al. [12] study formal statistical properties of sequence change and show how fluctuations follow a $-5/3$ Kolmogorov power and behave like an incremental Brownian process. In another study,

Wang et al. [10] investigate members of the family of β-Lactamases, enzymes involved in antibiotic resistance. In this study, they uncovered, via molecular simulations, important amino acid positions that share functional and dynamical features with another class of evolutionarily related proteins called Penicillin-binding proteins (PBP), enhancing our understanding of the dynamics of catalytic residues in the context of antibiotic resistance. In a third article, also concerned with the dynamics of protein evolution, Campitelli et al. [11] devise accurate metrics to quantify epistasis upon amino acid perturbations (EpiScore) and the asymmetric Dynamic Coupling Index (DCIasym) to measure how connected residues are affected depending on which residue has been perturbed. These metrics are relevant contributions to the study of allostery and the evolutionary forces that shape this important functional phenomenon.

In a final study, Sinner et al. [13] construct another information metric to predict the degree of specificity between molecules in two-component signaling networks. Molecular interactions between histidine kinases (HK) and response regulators (RR) have evolved towards amino acid specificity at the physical interface in the HK-RR complex where phosphotransfer occurs. A degree of coevolutionary strength at this interface can be quantified for a large number of organisms. The authors created a public web server called ELIHKSIR.org (Evolutionary Links Inferred for Histidine Kinase Sensors Interacting with Response regulators) to facilitate the prediction and analysis of these links and to assess the effect of mutations in interacting specificity.

All together, the methodological contributions presented in this issue of *Entropy* will help advance the study of molecular evolutionary dynamics through the lens of information theoretical metrics and a combination of structural modeling and molecular dynamics simulations.

Funding: The author's research is funded by the University of Texas at Dallas, NIH grant number R35GM133631, and NSF grant number MCB-1943442.

Acknowledgments: We acknowledge all author contributions to the Special Issue in Information Theory and Molecular Evolution: From Models to Structures and Dynamics.

Conflicts of Interest: The author declares no conflict of interest. The funders had no role in the design of the study; in the collection, analyses, or interpretation of data; in the writing of the manuscript, or in the decision to publish the results.

References

1. Fisher, R.A. XV.—The Correlation between Relatives on the Supposition of Mendelian Inheritance. *Trans. R. Soc. Edinb.* **1918**, *52*, 399–433. [CrossRef]
2. Grandy, W., Jr.; Milonni, P. *Physics and Probability: Essays in Honor of Edwin T. Jaynes*; Cambridge University Press: Cambridge, UK, 1993. [CrossRef]
3. Weigt, M.; White, R.A.; Szurmant, H.; Hoch, J.A.; Hwa, T. Identification of direct residue contacts in protein–protein interaction by message passing. *Proc. Natl. Acad. Sci. USA* **2009**, *106*, 67–72. [CrossRef] [PubMed]
4. Morcos, F.; Pagnani, A.; Lunt, B.; Bertolino, A.; Marks, D.S.; Sander, C.; Zecchina, R.; Onuchic, J.N.; Hwa, T.; Weigt, M. Direct coupling analysis of residue coevolution captures native contacts across many protein families. *Proc. Natl. Acad. Sci. USA* **2011**, *108*, E1293–E1301. [CrossRef]
5. Bitbol, A.F. Inferring interaction partners from protein sequences using mutual information. *PLoS Comput. Biol.* **2018**, *14*, e1006401. [CrossRef] [PubMed]
6. Marmier, G.; Weigt, M.; Bitbol, A.F. Phylogenetic correlations can suffice to infer protein partners from sequences. *PLoS Comput. Biol.* **2019**, *15*, e1007179. [CrossRef] [PubMed]
7. Hockenberry, A.J.; Wilke, C.O. Phylogenetic Weighting Does Little to Improve the Accuracy of Evolutionary Coupling Analyses. *Entropy* **2019**, *21*, 1000. [CrossRef] [PubMed]
8. Rodriguez Horta, E.; Barrat-Charlaix, P.; Weigt, M. Toward Inferring Potts Models for Phylogenetically Correlated Sequence Data. *Entropy* **2019**, *21*, 1090. [CrossRef]
9. Malinverni, D.; Barducci, A. Coevolutionary Analysis of Protein Subfamilies by Sequence Reweighting. *Entropy* **2019**, *21*, 1127. [CrossRef] [PubMed]
10. Wang, F.; Zhou, H.; Wang, X.; Tao, P. Dynamical Behavior of β-Lactamases and Penicillin-Binding Proteins in Different Functional States and Its Potential Role in Evolution. *Entropy* **2019**, *21*, 1130. [CrossRef]

1. Campitelli, P.; Ozkan, S.B. Allostery and Epistasis: Emergent Properties of Anisotropic Networks. *Entropy* **2020**, *22*, 667. [CrossRef] [PubMed]
2. Cadet, X.F.; Dehak, R.; Chin, S.P.; Bessafi, M. Non-Linear Dynamics Analysis of Protein Sequences. Application to CYP450. *Entropy* **2019**, *21*, 852. [CrossRef]
3. Sinner, C.; Ziegler, C.; Jung, Y.H.; Jiang, X.; Morcos, F. ELIHKSIR Web Server: Evolutionary Links Inferred for Histidine Kinase Sensors Interacting with Response Regulators. *Entropy* **2021**, *23*, 170. [CrossRef] [PubMed]

Article

Phylogenetic Weighting Does Little to Improve the Accuracy of Evolutionary Coupling Analyses

Adam J. Hockenberry * and **Claus O. Wilke**

Department of Integrative Biology, The University of Texas at Austin, Austin, TX 78712, USA;
wilke@austin.utexas.edu
* Correspondence: adam.hockenberry@utexas.edu

Received: 14 August 2019; Accepted: 10 October 2019; Published: 12 October 2019

Abstract: Homologous sequence alignments contain important information about the constraints that shape protein family evolution. Correlated changes between different residues, for instance, can be highly predictive of physical contacts within three-dimensional structures. Detecting such co-evolutionary signals via direct coupling analysis is particularly challenging given the shared phylogenetic history and uneven sampling of different lineages from which protein sequences are derived. Current best practices for mitigating such effects include sequence-identity-based weighting of input sequences and post-hoc re-scaling of evolutionary coupling scores. However, numerous weighting schemes have been previously developed for other applications, and it is unknown whether any of these schemes may better account for phylogenetic artifacts in evolutionary coupling analyses. Here, we show across a dataset of 150 diverse protein families that the current best practices out-perform several alternative sequence- and tree-based weighting methods. Nevertheless, we find that sequence weighting in general provides only a minor benefit relative to post-hoc transformations that re-scale the derived evolutionary couplings. While our findings do not rule out the possibility that an as-yet-untested weighting method may show improved results, the similar predictive accuracies that we observe across conceptually distinct weighting methods suggests that there may be little room for further improvement on top of existing strategies.

Keywords: direct coupling analysis; evolutionary coupling analysis; contact prediction; phylogenetic bias

1. Introduction

Correlated evolution of amino acid positions within a sequence alignment can be leveraged to inform structural models of proteins, predict mutational effects, and identify protein binding partners [1–5]. The ability to detect correlated evolution has been revolutionized by direct coupling analyses and other related methods that seek to re-construct one- and two-site marginal amino acid probabilities based on the observed distribution of sequence data [6–11]. Inference of two-site coupling parameters from a multiple sequence alignment is technically challenging, however, and numerous related approaches have been developed in recent years [9,10,12–17]. This intense focus on related methodologies stems from the fact that the highest scoring evolutionary coupling values are highly enriched in residue-residue pairs whose side-chains physically interact within three dimensional structures [18]. Evolutionary couplings can thus provide valuable information about structural constraints within and between protein families, while only requiring sequence information as inputs [15,19–22].

All methods to detect correlated evolution between different positions in a protein family require large numbers of representative sequences and therefore start by finding—and subsequently aligning—homologous sequences from large sequence databases [5]. An oft-remarked upon fact is that sequence databases are composed of a highly biased sample of life on Earth; some species are much more densely sampled than others (as are some genera, families, orders, etc.) [23–27]. Even if all extant life were equally well sampled and represented in sequence databases, species are related by complicated historical patterns and cannot be considered as independent observations [28].

Statistical issues arising from this shared phylogenetic history and biased sampling have long been noted by biologists [28]. The problem can be most clearly summarized by a toy example. In Figure 1A, we show a hypothetical sequence alignment and ask the question: What amino acid is preferred at the indicated site? At first glance, a phylogenetically agnostic method would simply count the frequency of different amino acids and conclude that valine (V, four occurrences) is preferred. However, accounting for phylogenetic relationships, a different perspective could reasonably conclude that threonine (T, three occurrences) is more highly preferred given that it occupies a substantially larger fraction of the phylogenetic tree and therefore dominates the evolutionary history of the protein family; the abundance of valines in the alignment is an apparent result of over-sampling one closely related lineage (which may represent numerous representatives of the same species, for example). Naively, the problem can be solved by simply selecting a single member from each species to prevent over-sampling. However, the issue remains equally problematic at other taxonomic levels (i.e., sampling numerous species from the same genus, numerous genera from the same family, etc.) and it is clear that a more general solution is required.

Prior research has shown that the best way to account for phylogenetic effects is to explicitly incorporate an evolutionary model into the statistical methods whenever possible [29–36]. However, this strategy can be challenging for certain problems [37] and simpler methods that differentially weight taxa according to their overall similarity to other taxa in a given dataset have been developed and applied for decades [38–46]. In the context of the toy example in Figure 1A, the choice of valine as the preferred amino acid comes from a model that weights each sequence uniformly. By down-weighting highly similar sequences, however, weighted frequencies could be used to come to the conclusion that threonine is instead the preferred amino acid. Instead of looking at preferred amino acid residues (one-site probabilities), evolutionary coupling analyses use sequence alignments to infer co-evolving positions via their two-site marginal probabilities. The current best practice for evolutionary coupling analyses is to down-weight sequences that are highly similar to one-another when inferring parameters from the multiple sequence alignment data. While this strategy appears in numerous methods, a systematic analysis of the benefit that sequence weighting provides in comparison to uniform weights, and an evaluation of different conceptually distinct strategies for assigning weights to sequences has not been performed to our knowledge.

Here, we evaluate existing weighting strategies alongside alternative tree- and sequence-based methods that have been proposed and used in various biological applications. We define the accuracy of a given method according to how well the resulting evolutionary couplings are able to predict residue–residue contacts within known representative structures of protein families [18]. Despite potential theoretical disadvantages, we find that the current best practice method of 80% sequence-identity-based weighting outperforms alternative methods that explicitly incorporate knowledge of phylogenetic relatedness. We show that a modification of this method provides a slight but insignificant improvement, and more broadly show that several methodologically distinct methods produce accuracies that are nearly indistinguishable both from one-another and from uniform weights.

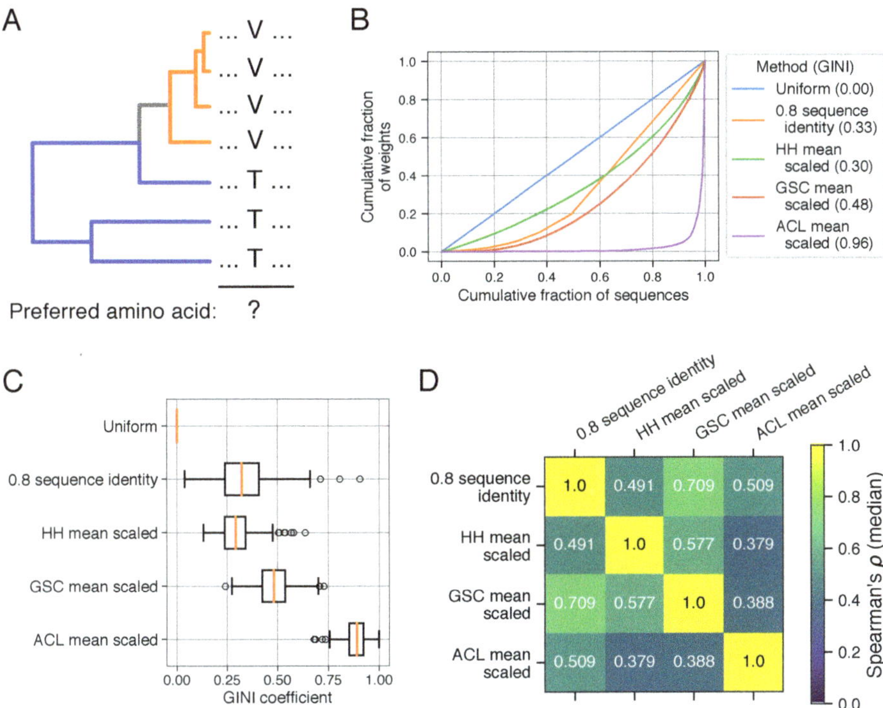

Figure 1. Weighting methods and their relationships in empirical datasets. (**A**) A toy example illustrating the problem of biased sampling and phylogenetic relatedness. Judging by their frequency (i.e., uniform weighting), valine (V) is the preferred amino acid at the indicated position. However, threonine (T) occupies a substantially larger proportion of the inferred evolutionary history. (**B**) For an example protein sequence alignment (PDB:1AOE), different weighting strategies produce a more- and less-uniform distribution of weights as visualized by the Lorenz curve. (**C**) The distribution of GINI coefficients for 150 protein families (higher coefficients correspond to a less uniform distribution of weights) using different weighting strategies (boxes span the 25th to 75th percentiles, and the red line indicates the median). (**D**) The median correlation coefficient (Spearman's ρ) of different weighting methods observed across the same 150 protein families.

2. Results

2.1. An Explanation of Weighting Methods

Many variants of evolutionary coupling analysis methods have been developed, and most methods implement a sequence-identity-based correction to mitigate the effect of phylogenetic relatedness [10,11,13]. Specifically, given n sequences in an alignment, the pairwise similarity of all sequences is calculated and the weight $W(i)$ of a given sequence i within an alignment equals the inverse of the total number of sequences j whose distance $d(i,j)$ to sequence i is less than some parameter λ:

$$W(i) = 1/ \sum_{j=1}^{n} I(i,j), \tag{1}$$

where n is the number of sequences in the alignment and $I(i,j)$ is an indicator variable defined as

$$I(i,j) = \begin{cases} 0 & \text{if } d_{i,j} < \lambda, \\ 1 & \text{if } d_{i,j} >= \lambda. \end{cases} \tag{2}$$

The distance $d(i,j)$ and the cutoff λ are usually measured as percent sequence identity: the number of identical residues between two aligned sequences divided by their total length.

Under this weighting scheme, highly unique sequences are given a weight value of 1, whereas sequences that are similar to others are assigned weights between 0 and 1 according to how many such similar sequences are in the alignment. Given this strategy, the effective number of sequences is simply the sum the weights assigned to all sequences, which takes a value between 0 and n.

Several possible issues arise from this weighting scheme. First, it is not immediately apparent what value of λ is most appropriate to use as a sequence identity threshold. While this parameter can be optimized for practical utility (the field has coalesced largely around a value of 80%), it is unclear what this value tells us about the co-evolutionary process or *why* it works so well. Second, this weighting scheme can produce some counter-intuitive results. Given an 80% sequence identity threshold, two otherwise independent sequences in an alignment sharing 99% sequence identity will each be assigned a weight of 0.5 reflecting their relative similarity to one another. In the same alignment, two sequences sharing 81% sequence identity will similarly each be assigned a weight of 0.5 despite being much more distinct from one another compared to the former pair. However, two sequences sharing 79% sequence identity will be assigned a weight of 1.0. Finally, the underlying phylogenetic history of the sequence evolution is ignored by this sequence-based comparison method which may inhibit its overall effectiveness.

Our goal here is not to exhaustively evaluate all possible strategies for assigning weights to sequences or tips on a phylogeny but rather to test several popular methods that represent logical starting points for possible improvements for use in evolutionary coupling analyses. Specifically, we decided to implement and test three algorithms: one sequence-based method and two conceptually distinct tree-based methods. The sequence-based method was proposed by Henikoff and Henikoff [44] and proceeds across each position by first awarding each observed residue at given position in an alignment an equal share of the weight for that position (where each position in the alignment has a starting weight of 1). The weights at that position for each sequence in the alignment are then assigned by dividing the weight assigned to each residue equally among all sequences sharing the same residue. Finally, the weight of a given sequence is simply the sum of the weights assigned to each position/residue. The method gives intuitively correct results for toy examples and has been used in numerous popular applications including HMMER and PSI-BLAST, with several different modifications for dealing with gap sequences [47,48].

We additionally implemented two tree-based methods that were initially proposed by Altschul et al. [38] (hereafter referred to as "ACL" weights) and Gerstein et al. [43] (hereafter referred to as "GSC" weights). The ACL method is equivalent to a model of electricity where a power source is plugged into the root of the tree, each branch provides resistance proportional to its length, and the current flowing out of each tip is used to determine the weights [38]. By contrast, the GSC method is a way of partitioning the branch lengths of a tree where the final weight of each tip is a weighted sum of all the branch lengths leading up to it [38,43]. Conceptually, ACL and GSC weights are quite distinct with GSC weights assigning a higher weight to tips that have particularly long branch lengths (and thus occupy a larger proportion of the tree) and ACL weights assigning the highest weights to sequences with particularly short branch lengths that reside closest to the root. We note that both metrics explicitly account for the underlying tree topology and thus require a previously constructed rooted evolutionary tree.

A notable caveat to the HH, ACL, and GSC weighting methods is that they do not provide intuitive *absolute* scales. The sum of all HH weights in their original formulation is equivalent to the length of the alignment, ACL weights are relative and sum to 1, and GSC weights are in units of branch length (substitutions per unit time) [38,43,44]. Differences in absolute scales will affect the output of co-evolutionary models because the regularization strength used during model fitting is proportional to the number of effective sequences. That is to say, a model fit to data where all weights are assigned a uniform value of 1 will be different from a model fit to the same set of sequences where all weights are assigned a uniform value of 0.1. Thus, both the relative differences in weights and their absolute scale are important considerations. For each of the three new methods, we employ two re-scaling strategies: First, we divide each weight value by the mean for that alignment, such that the weights for a given alignment will sum to n, where n is the number of sequences. Second, we divide each weight by the maximum observed weight in an alignment, such that the largest relative weight will be assigned a value of 1 and all other weights are some fraction of this.

For an example protein (PDB:1AOE), assigning weights to a sequence alignment/tree demonstrates that the methods vary substantially in how uniformly they distribute weights (Figure 1B). The GINI coefficient is a measurement of uniformity where values of zero correspond to uniform weights and values approaching 1 illustrate the case where a small number of sequences have very large weights while the remainder have very small weights. This relationship can be visualized by a Lorenz curve, which in this case plots the cumulative fraction of weights (y-axis) against the cumulative fraction of sequences (x-axis, sorted from lowest to highest weights). The Lorenz curves in Figure 1B show that ACL weights in particular result in a highly uneven distribution of weights. This finding holds more broadly across a dataset of 150 diverse protein families; the tree-based methods produce a more un-even distribution of weights, with ACL weights being particularly highly skewed (Figure 1C).

The different weighting schemes (when applied to the same multiple sequence alignment) are only modestly correlated with one-another. Figure 1D shows the median correlation (across the 150 protein families) observed among HH, GSC, and ACL as well as the commonly used 80% sequence-identity-based re-weighting method. In general, the weights produced by different methods on the same protein family are significantly positively correlated with one-another, but the correlations are fairly low, demonstrating that the weighting methods themselves are distinct. We additionally note that tree-based weighting methods may be subject to numerous errors during the tree construction and rooting process. We performed bootstrap resampling of multiple sequence alignments to compare the resulting weights to the originally calculated set of weights and found that they were significantly positively correlated (Figure S1). GSC weights, however, were much more robust (median Spearman's ρ of 0.84) compared to ACL weights (median Spearman's ρ of 0.61).

2.2. Sequence Weighting Does Little to Improve Contact Predictions

To test the effectiveness of different weighting methods, we calculated evolutionary couplings using the program CCMPredPy—a Python-based implementation of one of the most popular pseudo-likelihood based methods (CCMPred), which we modified to accept weights from externally supplied files—for 150 unique protein families with known structural representatives [13,16]. We next tested what fraction of the top L couplings for a given protein family (where L is the length of the reference sequence with a known three-dimensional structure) are true intramolecular residue–residue contacts—a metric known as the Positive Predictive Value (PPV) (see Section 4 for details) [18]. We separately quantified accuracies from the raw evolutionary couplings, entropy-corrected couplings, and Average Product Corrected (APC) couplings. The latter two post-hoc corrections have been shown to improve the accuracy of evolutionary couplings by accounting for uneven sequence entropies across positions in the alignment and perhaps the underlying phylogenetic structure [16,49].

As expected, we found that across all weighting schemes, the APC (and to a slightly lesser extent, the entropy-corrected) evolutionary couplings produce substantially more accurate results compared to raw coupling scores (Figure 2). In nearly all cases, sequence-identity-based weighting resulted in the highest accuracy. For the best performing APC coupling scores (Figure 2A), the commonly used λ parameter representing an 80% sequence identity threshold resulted in significantly higher accuracies compared to the uniform weight controls (Wilcoxon signed-rank test, $p < 0.001$). One phylogeny-based weighting method (GSC) and the HH sequence-based method were slightly more accurate than uniform weights provided that they were mean-scaled but the improvement was not significant in either case ($p = 0.09$ and $p = 0.1$, respectively); both methods were significantly less accurate than the 80% sequence-identity-based method ($p < 0.001$ for both cases). ACL weights by contrast generally performed poorly in all cases.

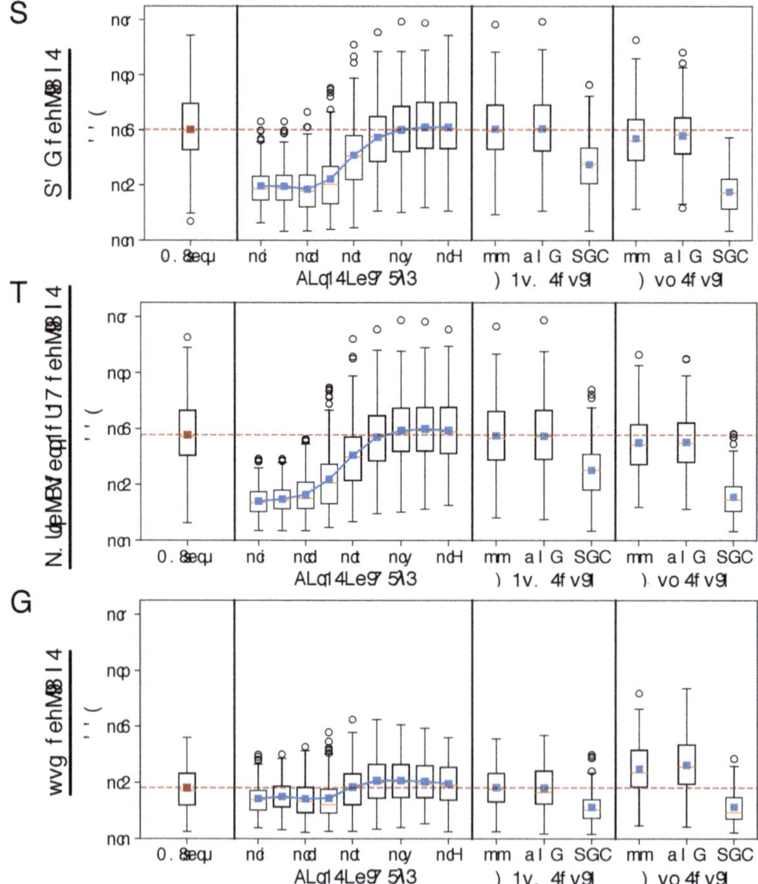

Figure 2. Testing the ability of evolutionary couplings to predict residue–residue contacts in representative structures. "Uniform" refers to the use of uniform weights for all sequences when fitting evolutionary coupling parameters (the red dashed line indicates the mean of this distribution and represents a baseline performance that methods should improve upon). "Threshold (λ)" refers to sequence-identity based weighting with different parameters, and "Mean scale" and "Max scale" refer to two different scalings of the indicated weighting methods (HH, GSC, and ACL). (**A**) Using APC couplings, the mean positive predictive values (PPVs) of the top L couplings vary across different weighting schemes used to infer evolutionary couplings. However, the only methods that significantly improve performance is sequence-identity-based re-weighting with $\lambda = 0.8$ or 0.9 (Wilcoxon signed-rank test, $p < 0.001$), but the magnitude of the improvement is modest (1.9% and 1.1% median improvement over uniform). (**B**) Using entropy-corrected evolutionary coupling values leads to similar conclusions that no weighting scheme substantially outperforms uniform weights. (**C**) Using raw evolutionary coupling values results in substantially higher accuracies for certain weighting methods relative to uniform, but the overall accuracies remain low compared to (**A**,**B**).

We note that even in the best case scenario the increase in PPV due to sequence weighting is comparatively small when compared to the large improvements in accuracy that result from the post-hoc APC and entropy corrections: median PPV for uniform weights are more than twice as high for APC couplings relative to raw couplings. Interestingly, the best performing weighting schemes substantially improve the accuracy of raw evolutionary couplings relative to the uniform weight control (Figure 2C, 44% median increase in PPV for max-scaled GSC weights, $p < 0.001$), but do comparatively little in the case of the more accurate APC couplings (Figure 2A, 2% median increase in PPV for 80% sequence-identity-based weights, $p < 0.001$).

A caveat noted above is that the regularization strength of the CCMPred model is proportional to the effective number of sequences. The typical value used for the pairwise regularization strength parameter ("LFACTOR") is 0.2, but this regularization strength was tuned for the previously best performing 80% sequence identity-based weights that are commonly employed. For the GSC, ACL, and HH methods, we tested a range of parameters (from 0.05 to 1.0) to see if a different pairwise regularization strength parameter might produce superior results (Figure S2). No combination of weighting method and parameter values, however, results in substantially improved accuracy for the best performing APC couplings. For all of the max scaled methods, smaller values of this parameter substantially improve results but only up to the level achieved by the best performing mean-scaled methods. Perhaps most notably, for entropy-corrected couplings, we found that larger pairwise regularization strength parameters were helpful for the best performing methods and brought the overall mean PPVs nearly on par with that of the APC couplings (Figure S2). The strength of regularization is thus an important consideration when evaluating different weighting schemes, but our finding that numerous methods achieve roughly the same overall accuracy remains unchanged from this analysis.

2.3. Weighting on Time-Scaled Trees

In Figure 1, we note that tree-based weighting methods produced a more un-even distribution of weights compared to the sequence-based weighting methods that we tested. A potential issue with both of the tree-based weighting methods that we consider here (aside from the possible noise/error in their calculation that was previously noted) is that the rates of evolution vary across phylogenetic trees and thus species are not equidistant from the root sequence. Phylogenetic trees reflect both the relationship between species and the rate of evolution along each branch. For trees consisting solely of extant species, numerous methods can re-scale trees to produce tips that are contemporaneous and equidistant from the root (Figure 3A) [50]. Since GSC and ACL weighting methods are significantly influenced by the overall distance from the root for individual tips, we reasoned that computing these weights on scaled-trees may produce less variable weights and perhaps more accurate results. We thus used the RelTime algorithm to transform each raw tree into a time-scaled tree and re-computed the weights for the two tree-based weighting methods on these RelTime trees [50].

For a given protein alignment, weights constructed in this manner display significantly less heterogeneity than weights calculated from the raw trees (Wilcoxon signed-rank test, $p < 0.001$). The PPVs of mean- and max-scaled weighting methods were significantly improved in all cases relative to weights computed on the raw trees (Figure 3B, results shown for APC couplings). The improvements were again comparatively small and no method out-performed 80% sequence-identity-based weights. However, PPVs with mean-scaled GSC weights calculated from RelTime trees were significantly higher than PPvs from uniform weighting (Wilcoxon signed-rank test, $p = 0.003$) and the difference in PPV between these weights and the best performing 80% sequence-identity-based weights was not significant ($p = 0.14$).

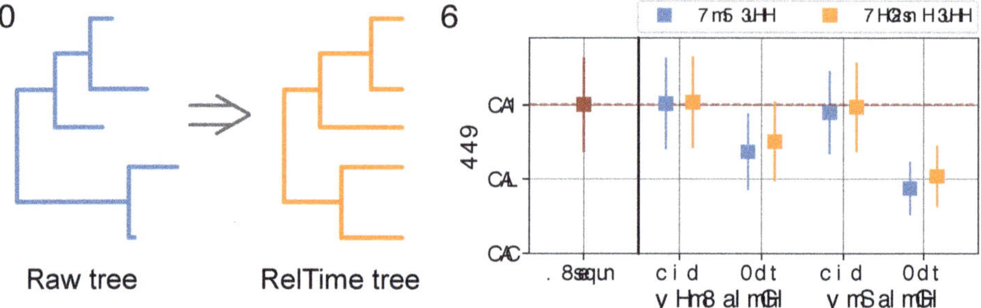

Figure 3. Tree re-scaling prior to calculation of weights slightly improves accuracies. (**A**) Raw, rooted phylogenetic trees can be converted to time-scaled trees with contemporaneous tips using the RelTime algorithm. (**B**) Sequence weights calculated from RelTime trees result in slightly better residue–residue contact prediction for the two tree-based weighting methods that we consider (and the two separate scalings of those weights). Shown is the mean PPV for 150 protein families using APC couplings, with error bars showing the standard deviation.

2.4. An Altered Sequence-Identity-Based Method That Accounts for Sequence Similarity

Thus far, we have shown that the current best practice of using sequence-identity-based weighting within a 80% sequence similarity neighborhood results in evolutionary couplings that have the highest power to predict intra-molecular residue–residue contacts. However, we also discussed some potentially counter-intuitive properties of this sequence-identity-based method. We thus developed and tested a variant of the sequence-identity-based method that down-weights sequences according to pairwise similarity and an identity threshold, but does so by accounting for the actual similarity between the sequences. Whereas the original method assigns each sequence a value of 1 and divides by the raw number of similar sequences (defined according to the λ parameter), our modification instead divides by the sum of a similarity-adjusted value for each sequence. Specifically,

$$W(i) = 1/\sum_{j=1}^{n} I_{\text{adj}}(i,j). \tag{3}$$

In contrast to Equation (2), $I_{\text{adj}}(i,j)$ produces a continuous range of values between 0 and 1:

$$I_{\text{adj}}(i,j) = \begin{cases} 0 & \text{if } d_{i,j} < \lambda, \\ (d_{i,j} - \lambda)/(1 - \lambda) & \text{if } d_{i,j} >= \lambda. \end{cases} \tag{4}$$

As in Equations (1) and (2), the distance $d_{i,j}$ and the cutoff λ are measured as percent sequence identity.

Using this method with a λ value of 0.8, two otherwise independent sequences in an alignment with 99% sequence identity will each be assigned a weight of 0.513 [or $1/(1 + 0.95)$, where $0.95 = (0.99 - 0.8)/(1 - 0.8)$], reflecting their high similarity to one another. In the same alignment, two sequences sharing only 81% sequence identity will by contrast each be assigned only a slightly decreased weight of 0.95 [or $1/(1 + 0.05)$, where $0.05 = (0.81 - 0.8)/(1 - 0.8)$]. All else being equal, the more similar sequences are, the more they will be down-weighted up to the given sequence identity threshold, at which point no further down-weighting occurs.

Comparing this similarity-adjusted sequence-identity-based method to the original method shows that the similarity-based adjustment produces more robust results across the range of possible values for λ (Figure 4). Across all of the different variants that we tested, similarity-adjusted sequence-identity-based weights with an identity parameter of 0.8 (and the APC, Figure 4A) produced evolutionary couplings with the highest median and mean PPV for the 150 protein families. PPVs resulting from this method were significantly higher than results from uniform weights (1.9% median and 3.7% mean increase in PPV, Wilcoxon signed-rank test $p < 0.001$) but the increase compared to 80% sequence-identity weights calculated in the original manner was slight and not significant (0% median and 0.3% mean increase in PPV, $p = 0.11$).

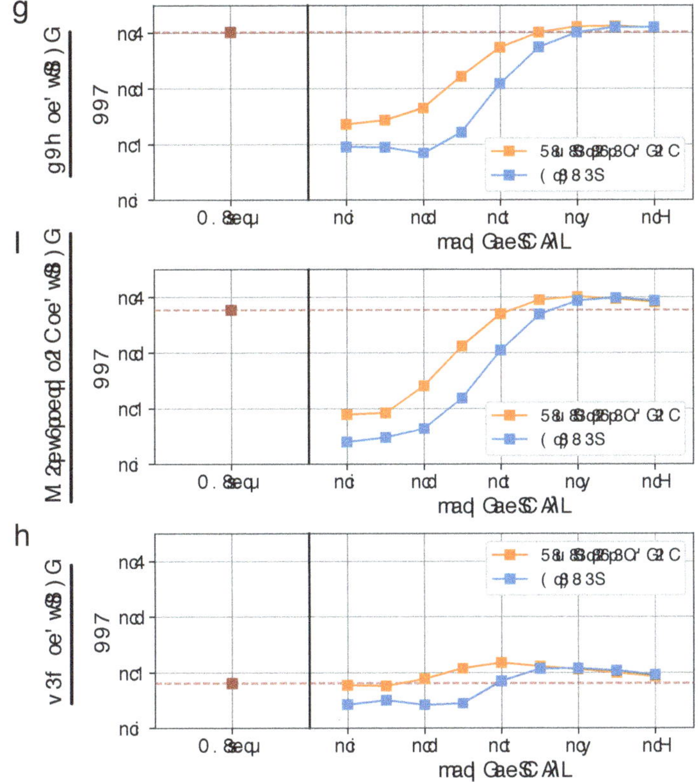

Figure 4. An altered sequence-identity-based method is more robust to parameter choice. (**A**) Using APC couplings, mean PPVs for similarity-adjusted sequence-identity-based weights are equal to or higher than PPVs calculated with the commonly used sequence-identity-based weights. (**B**) Same as in (**A**), using entropy-corrected evolutionary coupling values. (**C**) Same as in (**A,B**), using raw evolutionary coupling values.

3. Discussion

Natural sequence alignments are not composed of independently evolved lineages and instead have an unknown pattern of relationships that can be inferred and visualized as a phylogenetic tree. Statistical methods that fail to account for these relationships are expected to be biased, but in the case of direct coupling analyses a phylogenetically agnostic model has nevertheless proven valuable at predicting residue–residue contacts within protein structures [5,10,11]. Differential sequence weighting is commonly employed in such analyses as a way to partially mitigate phylogenetic effects, but the overall benefit that such weights provide has yet to be systematically interrogated. We have shown here that numerous (and conceptually distinct) weighting methods produce evolutionary couplings with a roughly equivalent ability to predict residue–residue contacts—given that the coupling values are transformed post-hoc via the average product correction (APC). We found that uniform, HH, GSC, and two variants of 80% sequence-identity-based weights all produce nearly indistinguishable accuracies from one another. While we have only evaluated a few different weighting methods and variants, the similar predictive power of top-performing weighting strategies (despite being substantially different from one-another, Figure 1D suggests that there may be little room for improvement on top of current best practices.

Intuitively, uneven sampling and phylogenetic biases are *expected* to introduce spurious effects into statistical models. Indeed, this is known to be the case in numerous contexts, such as when assessing the strength of correlations between discrete and continuous traits [28,34,36]. Nevertheless, we have shown here that using variable sequence weights to correct for these problems provides little (if any) practical benefit when attempting to predict residue–residue contacts. Why might this be the case? We caution that weights alone are an imperfect method of accounting for shared phylogenetic history, and in other contexts achieving accurate true and false positive rates from statistical tests requires more than simple re-weighting of data points [29,31,36,51,52].

In the context of evolutionary couplings, it is unclear whether uneven sampling and phylogenetic biases do not affect the fitting of coupling parameters as much as one might initially think, whether the APC (a post-hoc re-scaling procedure) largely corrects for any such factors, or whether weighting in general is simply an inadequate solution to the problem of phylogeny. Several lines of evidence currently indicate that the overall effect of phylogeny in direct coupling analysis models may be minimal. For instance, our results confirm previous findings showing that correcting for column-wise entropy produces comparable accuracies when compared to the average product correction (even though the latter is thought to partially correct for phylogenetic effects) [16]. Recent work has also shown that eigenvectors with the largest eigenvalues in a residue–residue covariance matrix strongly reflect the phylogenetic relatedness of the aligned sequences [53–55]. Removal of these eigenvectors substantially improves predictions of structural contacts in this conceptually distinct model, and variants of direct coupling analysis appear to achieve this same result via different means [55]. Future studies investigating the contribution of uneven sequence weighting towards these high value eigenvectors may be particularly illustrative about the impact of phylogenetic weights and their potential role in covariation analyses moving forward.

While we found that numerous weighting methods produce roughly equivalent end results on average, our findings raise other several potential issues that may be worthy of further study moving forward. We noted that many weighting methods do not clearly provide an intuitive absolute scale and instead assign weights to sequences (or tips in a phylogenetic tree) that are either relative or in irrelevant units. This can be problematic from a practical standpoint because most methods for inferring evolutionary coupling parameters between residue–residue pairs rely on some form of prior and the weight given to observed data relative to this prior may affect results. For the HH, GSC, and ACL methods, we found that two different scaling procedures (which maintain relative weights within a dataset but change their absolute values) produced varying accuracies (Figure 2). With the exception of star phylogenies, the effective sample size from phylogenetically structured data is strictly less than the number of sequences/data points analyzed. More accurately estimating the effective sample size and scaling weights accordingly may improve the performance of different weighting schemes beyond what we observed here.

Additionally, the HH, GSC, and ACL methods do not include a free parameter that can be tuned to improve results. We validated that an 80% sequence-identity neighborhood is optimal using the currently accepted method and a similarity-adjusted variant, but this 80% value is a free parameter that has been optimized to produce the highest accuracy for sequence-identity-based weighting. What we believe the optimality of this parameter represents in practice is that once two sequences diverge past approximately 80% similarity, their evolution is effectively independent. If this is the case, down-weighting sequences that for instance share 50% sequence identity might make little sense. Fully answering this question, however, would require testing a range of regularization strengths since the effective number of sequences at that level of sequence identity down-weighting is substantially less than at 80% sequence identity. By contrast, the HH, GSC, and ACL methods all inherently compare each sequence to every other sequence in a global manner. It seems possible that some phylogenetic tree transformation may be able to introduce the same intuition of ignoring evolutionary relatedness past some threshold level into tree-based weighting methods [30,32]. The best way to perform such re-scaling, or how to perform something conceptually similar for HH weights, is a promising area for future research.

Another potential area for future research is to specifically investigate the cases when sequence re-weighting makes the largest/smallest impact on PPV relative to uniform weights. The overall predictive power of evolutionary couplings varies both within and between protein families. Previous work has shown that within an individual protein family, structural contacts that are mediated by side-chain interactions are most likely to be detected by co-evolutionary methods, as opposed to those mediated by atomic interactions in the peptide backbone [18]. Between protein families, the substantial variability in PPVs result from numerous factors including but not limited to the number of sequences in an alignment, the diversity of those sequences, homo-oligomerization, alignment errors from repeat proteins, and family structural variation [56]. While we found that PPVs are highly variable across protein families and that sequence re-weighting can help increase these scores a relatively small amount on average (Figure 2), the magnitude of this increase is higher for some families compared to others. Being able to associate the magnitude of the increase with properties of the sequences or tree, such as their diversity or bias, may provide interesting clues about the general ineffectiveness of sequence weighting or insight into novel strategies that could better account for these effects.

Despite being weakly correlated with one another, uniform, 80% sequence identity, HH, and GSC weights perform roughly equivalently at predicting residue–residue contacts. We recommend that any method with substantially improved performance should become the standard but computational complexity and run-time are real constraints that most researchers should additionally consider. Once a phylogenetic tree is constructed, the cost of calculating the different weights that we considered here is negligible relative to the run-time of inference algorithms. However, given the ideal size of protein family alignments (thousands to tens of thousands of sequences), the most accurate methods for phylogenetic tree construction are computationally infeasible. Even more rapid methods, such as those we employed here, may substantially increase the overall run-time for a pipeline relying on tree-based weights. At present, our current results give no indication that the increased computational time and complexity of tree-construction will provide any benefit. If we were to see such a benefit, the choice of whether a few percent increase in accuracy would be worth doubling (or worse) the run-time for a protein family of interest would be dependent on the researcher and the application.

While several methods were nearly indistinguishable from one another in terms of their accuracies, we did find that a slightly modified sequence-identity-based re-weighting method that accounts for sequence similarity actually performs the best of any method that we tested. This method does not require calculation of a phylogenetic tree and therefore has an overall run-time that is virtually equivalent assuming uniform weights or existing best practices. However, using either the original or similarity-adjusted sequence-identity-based weighting can be expected to offer less than a few percent improvement in accuracy compared to uniform weights, which completely ignore phylogeny. We therefore speculate that if phylogenetic effects are truly problematic for inferring co-evolution—and we caution that this is not necessarily a given—then substantial improvements to existing methods may require the explicit incorporation of phylogenies and time-dependent sequence evolution rather than heuristic re-weighting strategies.

4. Materials and Methods

4.1. Description of the Dataset

For all of our analyses, we used the so-called "psicov" dataset—an existing set of 150 distinct protein structures with corresponding multiple sequence alignments that have been used in numerous benchmark studies for predicting residue–residue contacts from evolutionary couplings [14,57,58].

All sequence and structure data were taken directly from Jones and Kandathil [58], but, given the large number of different analyses that we ran, we first randomly down-sampled each alignment to a maximum of 1001 sequences (1000 sequences plus the mandated inclusion of the reference protein sequence).

4.2. Phylogenetic Tree Construction

For each sequence alignment in our dataset, we constructed a rough phylogenetic tree using the double precision version of FastTree2 (v.2.1.10; LG model, gamma distributed rate variation, with pseudocounts) [59]. We next adjusted the branch lengths on each guide tree by running the alignment and the template tree through the more accurate IQtree software (v1.6.9; LG model, Gamma-distributed rate variation with 20 categories) [60]. Finally, we rooted the resulting trees using the mid-point method [61].

For RelTime trees, we implemented our own version of the RelTime algorithm as described in the original manuscript while ensuring that our method produced similar results [50]. We note here only that our implementation does not perform a statistical test (and subsequent alteration of rates) at the end of the algorithm to ensure that rate changes are significant.

4.3. Weighting Methods

We developed all of our weighting methods from scratch using custom python programs that heavily leveraged tools from the Biopython package [61]. For sequence identity weighting and the novel similarity-adjusted version we propose here, details are presented in the main text, Equations (1)–(4). We ensured that our own version of sequence-identity-based weighting was equivalent to the method implemented within CCMpredPy by comparing the resulting effective number of sequences metrics and accuracies and finding them to be identical.

For HH based weights, we followed the procedure outlined in the initial paper and ensured that our implementation gave the desired results on the toy examples presented therein [44]. Researchers have pointed out subsequent modifications to this method [47,48] concerning how to effectively treat gap sequences. Rather than treating these as a 21st character as some implementations have done, our implementation assigns gap sequences a weight value of zero. Further, each column in the alignment is weighted from 0 to 1 according to the fraction of non-gapped positions. In this manner, alignment positions with more gaps are assigned lower weights and the positions with gaps themselves contribute a weight of zero. Summation and calculation of final weights follows the published procedure [44]. However, since the units and absolute value of these weights are not intuitive, we finally re-scaled the weights via separate mean- and max-scaling procedures. In mean-scaling, we calculate the mean of all weights determined via the HH algorithm for a particular sequence alignment and then divide the weight of each sequence in the alignment by this value. This ensures that the sum of all final weights will be equal to the number of sequences in the alignment (n). In the separate max-scaling procedure, we find the maximum weight observed for a particular sequence alignment, and subsequently divide all weights in the alignment by this value. The sum of all weights following this procedure is guaranteed to be some value less than the total number of sequences (n).

For ACL and GSC weights, we again followed the procedures outlined in the respective manuscripts [38,43] and ensured that our implementations produced identical results to the examples presented therein. As with HH, calculation of final weights occurred by (separately) scaling the weight values via their mean and maximum values as noted above.

4.4. Evolutionary Coupling Analysis

We chose to use CCMpredPy (v1.0.0, contained as part of the CCMgen package) [13,16] for all evolutionary coupling analyses since we were able to modify the source code for this popular method to accept externally supplied weights in the form of a simple text file where the weight value for each sequence corresponded to its line in the input sequence file. We used the default values with the `ofn-pll` flag corresponding to the pseudo-likelihood optimization of coupling parameters. For each different weighting method that we tested, we outputted files corresponding to the raw, entropy-corrected, and average product corrected coupling matrices.

4.5. Structural Analysis and Accuracy Determination

We used the `.PDB` files provided as part of the psicov dataset and for each structure computed a matrix of residue–residue distances. Each distance value is measured according to the geometric center for all side-chain heavy atoms for a particular residue (including the $C\beta$ atom, excluding the $C\alpha$ atom) [18]. In the case of glutamine, the side-chain center coordinates were assigned to the $C\alpha$ atom. We determined residue–residue contacts according to a uniform 7.5 angstrom threshold for all proteins.

We determined the accuracy of evolutionary couplings by determining how well they were able to predict residue–residue contacts within a reference structure. We first selected the top L-ranked couplings for each dataset, where L corresponds to the length of the reference protein sequence (i.e., the sequence for which we have a known structure). The PPV for a particular dataset corresponds to the fraction of those top L-ranked couplings that are classified as residue–residue contacts according to the above definition.

Supplementary Materials: The following are available online at http://www.mdpi.com/1099-4300/21/10/1000/s1.

Author Contributions: Conceptualization, A.J.H. and C.O.W.; methodology, A.J.H. and C.O.W.; software, A.J.H.; validation, A.J.H. and C.O.W.; formal analysis, A.J.H.; investigation, A.J.H.; resources, A.J.H. and C.O.W.; data curation, A.J.H.; writing—original draft preparation, A.J.H.; writing—review and editing, A.J.H. and C.O.W.; visualization, A.J.H. and C.O.W.; supervision, C.O.W.; project administration, C.O.W.; and funding acquisition, A.J.H. and C.O.W.

Funding: This work was funded by National Institutes of Health grant R01 GM088344 and by F32 GM130113.

Conflicts of Interest: The authors declare no conflict of interest. The funders had no role in the design of the study; in the collection, analyses, or interpretation of data; in the writing of the manuscript, or in the decision to publish the results.

Abbreviations

The following abbreviations are used in this manuscript:

HH	weights derived via the method of Henikoff and Henikoff [44]
GSC	weights derived via the method of Gerstein et al. [43]
ACL	weights derived via the method of Altschul et al. [38]
APC	Average Product Correction/ed
PPV	Positive Predictive Value

References

1. Gobel, U.; Sander, C.; Schneider, R.; Valencia, A. Correlated Mutations and Residue Contacts in Proteins. *Proteins* **1994**, *18*, 309–317. [CrossRef] [PubMed]
2. Hopf, T.A.; Scharfe, C.P.I.; Rodrigues, J.P.G.L.M.; Green, A.G.; Kohlbacher, O.; Sander, C.; Bonvin, A.M.J.J.; Marks, D.S. Sequence co-evolution gives 3D contacts and structures of protein complexes. *eLife* **2014**, *3*, 1–45. [CrossRef] [PubMed]
3. Hopf, T.A.; Ingraham, J.B.; Poelwijk, F.J.; Schärfe, C.P.; Springer, M.; Sander, C.; Marks, D.S. Mutation effects predicted from sequence co-variation. *Nat. Biotechnol.* **2017**, *35*, 128–135. [CrossRef] [PubMed]
4. Kamisetty, H.; Ovchinnikov, S.; Baker, D. Assessing the utility of coevolution-based residue-residue contact predictions in a sequence- and structure-rich era. *Proc. Natl. Acad. Sci. USA* **2013**, *110*, 15674–15679. [CrossRef]
5. Ovchinnikov, S.; Park, H.; Varghese, N.; Huang, P.S.; Pavlopoulos, G.A.; Kim, D.E.; Kamisetty, H.; Kyrpides, N.C.; Baker, D. Protein structure determination using metagenome sequence data. *Science* **2017**, *355*, 294–298. [CrossRef]

6. Lapedes, A.S.; Giraud, B.G.; Liu, L.; Stormo, G.D. Correlated mutations in models of protein sequences: Phylogenetic and structural effects. *Stat. Mol. Biol. Genet.* **1999**, *33*, 236–256. [CrossRef]

7. Burger, L.; Van Nimwegen, E. Accurate prediction of protein-protein interactions from sequence alignments using a Bayesian method. *Mol. Syst. Biol.* **2008**, *4*. [CrossRef]

8. Weigt, M.; White, R.A.; Szurmant, H.; Hoch, J.A.; Hwa, T. Identification of direct residue contacts in protein-protein interaction by message passing. *Proc. Natl. Acad. Sci. USA* **2009**, *106*, 67–72. [CrossRef]

9. Burger, L.; Van Nimwegen, E. Disentangling direct from indirect co-evolution of residues in protein alignments. *PLoS Comput. Biol.* **2010**, *6*. [CrossRef]

10. Marks, D.S.; Colwell, L.J.; Sheridan, R.; Hopf, T.A.; Pagnani, A.; Zecchina, R.; Sander, C. Protein 3D structure computed from evolutionary sequence variation. *PLoS ONE* **2011**, *6*. [CrossRef]

11. Morcos, F.; Pagnani, A.; Lunt, B.; Bertolino, A.; Marks, D.S.; Sander, C.; Zecchina, R.; Onuchic, J.N.; Hwa, T.; Weigt, M. Direct-coupling analysis of residue coevolution captures native contacts across many protein families. *Proc. Natl. Acad. Sci. USA* **2011**, *108*, E1293–E1301. [CrossRef] [PubMed]

12. Ekeberg, M.; Lövkvist, C.; Lan, Y.; Weigt, M.; Aurell, E. Improved contact prediction in proteins: Using pseudolikelihoods to infer Potts models. *Phys. Rev. E* **2013**, *87*, 1–16. [CrossRef] [PubMed]

13. Seemayer, S.; Gruber, M.; Söding, J. CCMpred - Fast and precise prediction of protein residue-residue contacts from correlated mutations. *Bioinformatics* **2014**, *30*, 3128–3130. [CrossRef] [PubMed]

14. Jones, D.T.; Singh, T.; Kosciolek, T.; Tetchner, S. MetaPSICOV: Combining coevolution methods for accurate prediction of contacts and long range hydrogen bonding in proteins. *Bioinformatics* **2015**, *31*, 999–1006. [CrossRef]

15. Figliuzzi, M.; Barrat-Charlaix, P.; Weigt, M. How pairwise coevolutionary models capture the collective residue variability in proteins? *Mol. Biol. Evol.* **2018**, *35*, 1018–1027. [CrossRef]

16. Vorberg, S.; Seemayer, S.; Söding, J. Synthetic protein alignments by CCMgen quantify noise in residue-residue contact prediction. *PLoS Comput. Biol.* **2018**, *14*, e1006526. [CrossRef]

17. Hopf, T.A.; Green, A.G.; Schubert, B.; Mersmann, S.; Schärfe, C.P.; Ingraham, J.B.; Toth-Petroczy, A.; Brock, K.; Riesselman, A.J.; Palmedo, P.; et al. The EVcouplings Python framework for coevolutionary sequence analysis Thomas. *Bioinformatics* **2018**, *35*, 1582–1584. [CrossRef]

18. Hockenberry, A.J.; Wilke, C.O. Evolutionary couplings detect side-chain interactions. *PeerJ* **2019**, *e7280*, 1–22. [CrossRef]

19. Morcos, F.; Jana, B.; Hwa, T.; Onuchic, J.N. Coevolutionary signals across protein lineages help capture multiple protein conformations. *Proc. Natl. Acad. Sci. USA* **2013**, *110*, 20533–20538. [CrossRef]

20. Bitbol, A.F.; Dwyer, R.S.; Colwell, L.J.; Wingreen, N.S. Inferring interaction partners from protein sequences. *Proc. Natl. Acad. Sci. USA* **2016**, *113*, 12180–12185. [CrossRef]

21. Uguzzoni, G.; John Lovis, S.; Oteri, F.; Schug, A.; Szurmant, H.; Weigt, M. Large-scale identification of coevolution signals across homo-oligomeric protein interfaces by direct coupling analysis. *Proc. Natl. Acad. Sci. USA* **2017**, *114*, E2662–E2671. [CrossRef] [PubMed]

22. Cong, Q.; Anishchenko, I.; Ovchinnikov, S.; Baker, D. Protein interaction networks revealed by proteome coevolution. *Science* **2019**, *365*, 185–189. [CrossRef] [PubMed]

23. Bonnet, X.; Shine, R.; Lourdais, O. Taxonomic chauvinism. *Trends Ecol. Evol.* **2002**, *17*, 1–3. [CrossRef]

24. Chen, C.; Natale, D.A.; Finn, R.D.; Huang, H.; Zhang, J.; Wu, C.H.; Mazumder, R. Representative Proteomes: A Stable, Scalable and Unbiased proteome set for sequence analysis and functional annotation. *PLoS ONE* **2011**, *6*, e18910. [CrossRef] [PubMed]

25. Rinke, C.; Schwientek, P.; Sczyrba, A.; Ivanova, N.N.; Anderson, I.J.; Cheng, J.F.; Darling, A.; Malfatti, S.; Swan, B.K.; Gies, E.A.; et al. Insights into the phylogeny and coding potential of microbial dark matter. *Nature* **2013**, *499*, 431–437. [CrossRef]

26. Troudet, J.; Grandcolas, P.; Blin, A.; Vignes-Lebbe, R.; Legendre, F. Taxonomic bias in biodiversity data and societal preferences. *Sci. Rep.* **2017**, *7*, 1–14. [CrossRef]

27. Titley, M.A.; Snaddon, J.L.; Turner, E.C. Scientific research on animal biodiversity is systematically biased towards vertebrates and temperate regions. *PLoS ONE* **2017**, *12*, 1–14. [CrossRef]

28. Felsenstein, J. Phylogenies and the comparative method. *Am. Nat.* **1985**, *125*, 1–15. [CrossRef]

29. Grafen, A. The phylogenetic regression. *Philos. Trans. R. Soc. B* **1989**, *326*, 119–157. [CrossRef]

30. Pagel, M. Inferring historical patterns of biological evolution. *Nature* **1999**, *401*, 877–884. [CrossRef]

31. Rohlf, F.J. Comparative methods for the analysis of continuous variables: geometric interpretations. *Evolution* **2001**, *55*, 2143–2160. [CrossRef] [PubMed]

32. Blomberg, S.P.; Garland, T., Jr.; Ives, A.R. Testing for phylogenetic signal in comparative data: Behavioral traits are more labile. *Evolution* **2003**, *57*, 717–745. [CrossRef] [PubMed]

33. Ives, A.R.; Midford, P.E.; Garland, T., Jr. Within-species variation and measurement error in phylogenetic comparative methods. *Syst. Biol.* **2007**, *56*, 252–270. [CrossRef] [PubMed]

34. Ives, A.R.; Garland, T., Jr. Phylogenetic Regression for Binary Dependent Variables. *Syst. Biol.* **2010**, *59*, 9–26._9. [CrossRef] [PubMed]

35. Revell, L.J. Size-correction and principal components for interspecific comparative studies. *Evolution* **2009**, *63*, 3258–3268. [CrossRef] [PubMed]

36. Revell, L.J. Phylogenetic signal and linear regression on species data. *Methods Ecol. Evol.* **2010**, *1*, 319–329. [CrossRef]

37. Uyeda, J.C.; Zenil-Ferguson, R.; Pennell, M.W. Rethinking phylogenetic comparative methods. *Syst. Biol.* **2018**, *67*, 1091–1109. [CrossRef]

38. Altschul, S.F.; Carroll, R.J.; Lipman, D.J. Weights for data related by a tree. *J. Mol. Biol.* **1989**, *207*, 647–653. [CrossRef]

39. Vingron, M.; Argos, P. A fast and multiple sequence alignment algorithm. *Bioinformatics* **1989**, *5*, 115–121. [CrossRef]

40. Sibbald, P.R.; Argos, P. Weighting aligned protein or nucleic acid sequences to correct for unequal representation. *J. Mol. Biol.* **1990**, *216*, 813–818. [CrossRef]

41. Vingron, M.; Sibbald, P.R. Weighting in sequence space: A comparison of methods in terms of generalized sequences. *Proc. Natl. Acad. Sci. USA* **1993**, *90*, 8777–8781. [CrossRef] [PubMed]

42. Thompson, J.D.; Higgins, D.G.; Gibson, T.J. Improved sensitivity of profile searches through the use of sequence weights and gap excision. *Bioinformatics* **1994**, *10*, 19–29. [CrossRef]

43. Gerstein, M.; Sonnhammer, E.L.; Chothia, C. Volume changes in protein evolution. *J. Mol. Biol.* **1994**, *236*, 1067–1078. [CrossRef]

44. Henikoff, S.; Henikoff, J.G. Position-based sequence weights. *J. Mol. Biol.* **1994**, *243*, 574–578. [CrossRef]

45. Krogh, A.; Mitchison, G. Maximum entropy weighting of aligned sequences of proteins or DNA. *Proc. Int. Conf. Intell. Syst. Mol. Biol.* **1995**, *3*, 215–21. [PubMed]

46. Stone, E.A.; Sidow, A. Constructing a meaningful evolutionary average at the phylogenetic center of mass. *BMC Bioinform.* **2007**, *8*, 1–13. [CrossRef] [PubMed]

47. Altschul, S.F.; Madden, T.L.; Schäffer, A.A.; Zhang, J.; Zhang, Z.; Miller, W.; Lipman, D.J. Gapped BLAST and PSI-BLAST: A new generation of protein database search programs. *Nucleic Acids Res.* **1997**, *25*, 3389–3402. [CrossRef] [PubMed]

48. Eddy, S.R. Profile hidden Markov models. *Bioinformatics* **1998**, *14*, 755–763. [CrossRef]

49. Dunn, S.D.; Wahl, L.M.; Gloor, G.B. Mutual information without the influence of phylogeny or entropy dramatically improves residue contact prediction. *Bioinformatics* **2008**, *24*, 333–340. [CrossRef]

50. Tamura, K.; Battistuzzi, F.U.; Billing-Ross, P.; Murillo, O.; Filipski, A.; Kumar, S. Estimating divergence times in large molecular phylogenies. *Proc. Natl. Acad. Sci. USA* **2012**, *109*, 19333–19338. [CrossRef]

51. Bruno, W.J. Modeling residue usage in aligned protein sequences via maximum likelihood. *Mol. Biol. Evol.* **1996**, *13*, 1368–1374. [CrossRef] [PubMed]

52. Newberg, L.A.; McCue, L.A.; Lawrence, C.E. The Relative Inefficiency of Sequence Weights Approaches in Determining a Nucleotide Position Weight Matrix. *Stat. Appl. Genet. Mol. Biol.* **2005**, *4*. [CrossRef] [PubMed]

53. Patterson, N.; Price, A.L.; Reich, D. Population Structure and Eigenanalysis. *PLoS Genet.* **2006**, *2*. [CrossRef] [PubMed]

54. Cocco, S.; Monasson, R.; Weigt, M. From Principal Component to Direct Coupling Analysis of Coevolution in Proteins: Low-Eigenvalue Modes are Needed for Structure Prediction. *PLoS Comput. Biol.* **2013**, *9*. [CrossRef]

55. Qin, C.; Colwell, L.J. Power law tails in phylogenetic systems. *Proc. Natl. Acad. Sci. USA* **2018**, *115*, 690–695. [CrossRef]

56. Anishchenko, I.; Ovchinnikov, S.; Kamisetty, H.; Baker, D. Origins of coevolution between residues distant in protein 3D structures. *Proc. Natl. Acad. Sci. USA* **2017**. [CrossRef]

Entropy **2019**, *21*, 1000

57. Jones, D.T.; Buchan, D.W.; Cozzetto, D.; Pontil, M. PSICOV: Precise structural contact prediction using sparse inverse covariance estimation on large multiple sequence alignments. *Bioinformatics* **2012**, *28*, 184–190. [CrossRef]

58. Jones, D.T.; Kandathil, S.M. High precision in protein contact prediction using fully convolutional neural networks and minimal sequence features. *Bioinformatics* **2018**, *34*, 3308–3315. [CrossRef]

59. Price, M.N.; Dehal, P.S.; Arkin, A.P. FastTree 2 - Approximately maximum-likelihood trees for large alignments. *PLoS ONE* **2010**, *5*. [CrossRef]

60. Nguyen, L.T.; Schmidt, H.A.; Von Haeseler, A.; Minh, B.Q. IQ-TREE: A fast and effective stochastic algorithm for estimating maximum-likelihood phylogenies. *Mol. Biol. Evol.* **2015**, *32*, 268–274. [CrossRef]

61. Cock, P.J.; Antao, T.; Chang, J.T.; Chapman, B.A.; Cox, C.J.; Dalke, A.; Friedberg, I.; Hamelryck, T.; Kauff, F.; Wilczynski, B.; et al. Biopython: Freely available Python tools for computational molecular biology and bioinformatics. *Bioinformatics* **2009**, *25*, 1422–1423. [CrossRef] [PubMed]

Sample Availability: Raw and processed data used in this manuscript have been deposited at: https://doi.org/10.5281/zenodo.3368652. All necessary code to replicate the analyses presented here have been deposited at: https://github.com/adamhockenberry/dca-weighting.

Article

Toward Inferring Potts Models for Phylogenetically Correlated Sequence Data

Edwin Rodriguez Horta [1,2] , **Pierre Barrat-Charlaix** [1,3] **and Martin Weigt** [1,*]

1   Laboratoire de Biologie Computationnelle et Quantitative (LCQB), Institut de Biologie Paris-Seine, Sorbonne Université, Centre national de la recherche scientifique (CNRS), 75005 Paris, France; erodriguezh1990@gmail.com (E.R.H.); p.barrat@live.fr (P.B.-C.)

2   Group of Complex Systems and Statistical Physics, Department of Theoretical Physics, Physics Faculty, University of Havana, La Habana 10400, Cuba

3   Biozentrum, University of Basel, 4056 Basel, Switzerland

*   Correspondence: martin.weigt@upmc.fr

Received: 25 September 2019; Accepted: 6 November 2019; Published: 7 November 2019

Abstract: Global coevolutionary models of protein families have become increasingly popular due to their capacity to predict residue–residue contacts from sequence information, but also to predict fitness effects of amino acid substitutions or to infer protein–protein interactions. The central idea in these models is to construct a probability distribution, a Potts model, that reproduces single and pairwise frequencies of amino acids found in natural sequences of the protein family. This approach treats sequences from the family as independent samples, completely ignoring phylogenetic relations between them. This simplification is known to lead to potentially biased estimates of the parameters of the model, decreasing their biological relevance. Current workarounds for this problem, such as reweighting sequences, are poorly understood and not principled. Here, we propose an inference scheme that takes the phylogeny of a protein family into account in order to correct biases in estimating the frequencies of amino acids. Using artificial data, we show that a Potts model inferred using these corrected frequencies performs better in predicting contacts and fitness effect of mutations. First, only partially successful tests on real protein data are presented, too.

Keywords: phylogeny; co-evolution; direct coupling analysis

---

1. Introduction

Based on the rapidly growing availability of biological sequence data [1–3], statistical models of sequences have gained considerable interest over the last years [4–7]. In this context, the direct coupling analysis (DCA) [8] takes inspiration from inverse statistical physics [9]: it aims at describing the sequence variability of sets of evolutionarily related protein sequences—so-called homologous protein families—via Potts models. Such a model gives a probability

$$P(\underline{A}) = \frac{1}{Z} \exp\left\{ \sum_{1 \leq i < j \leq L} J_{ij}(A_i, A_j) + \sum_{1 \leq i \leq L} h_i(A_i) \right\} \qquad (1)$$

to each aligned amino acid sequence $\underline{A} = (A_1, ..., A_L)$ of length L, with the $A_i \in \mathcal{A} = \{A, C, ..., Y, -\}$ being either one of the 20 amino acids, or an alignment gap, "$-$", representing amino acid insertions or deletions. The total alphabet size is $q = |\mathcal{A}| = 21$. Strong statistical couplings J_{ij} between different positions have been found to be indicative of contacts of the corresponding amino acids in the three-dimensional protein fold, thereby facilitating protein structure prediction from sequence information [10,11]. Furthermore, the statistical energy landscape (i.e., the Hamiltonian $\mathcal{H}(\underline{A}) =$

$-\sum_{1 \le i < j \le L} J_{ij}(A_i, A_j) - \sum_{1 \le i \le L} h_i(A_i)$ of the Potts model in Equation (1)) around a sequence has been found to be informative about the effects of mutations on the protein's functionality (or fitness) [12].

Sequence data for protein families are typically available as multiple-sequence alignments (MSA), i.e., as collections $\{\underline{A}^m\}_{m=1...M}$ of M distinct sequences of the same (aligned) length L. To fit the model $P(\underline{A})$ in Equation (1) to these data, typically a very strong assumption is made: the MSA is considered an independently and identically distributed sample of the statistical model. This implies that the model can be inferred by maximizing the likelihood

$$\mathcal{L}_{i.i.d.}(\{J_{ij}(A, B), h_i(A)\} \,|\, \{\underline{A}^m\}) = \prod_{m=1}^{M} P(\underline{A}^m) \tag{2}$$

over all couplings $J_{ij}(A, B)$ and fields $h_i(A)$. Although this task is computationally hard—it requires in particular the calculation of the partition function Z in Equation (1) as a sum over 21^L sequences—numerous approximation schemes have been developed and are reviewed in [7,9].

However, the evolutionary history of proteins is in evident contradiction with the assumption of statistical independence between sequences. The very notion of homologous protein families implies that present sequences derive from a common ancestor. Even if the divergence time from this common ancestor is long enough to result in overall high sequence diversity, some protein sequences may be found in closely related species, or may go back to a relatively recent event of duplication or horizontal gene transfer. This is commonly observable in MSA, where sequences differing by only few amino acids are frequent.

The evolutionary history of a protein family is typically represented by a phylogenetic tree [13], cf. Figure 1 for a simple example. Sequences observable today correspond to the leaves of this tree, and the common ancestor to its root. Branching points correspond to events separating two sequences, typically via speciation, duplication, or horizontal gene transfer. On distinct branches, proteins are assumed to evolve independently.

Figure 1. Homologous proteins constituting a multiple-sequence alignment (MSA) are related by common ancestors through a phylogenetic tree.

However, if the branching event separating two sequences, $\underline{A}^1$ and $\underline{A}^2$, took place some time Δt in the past, the joint probability should be written as $P(\underline{A}^1, \underline{A}^2 | \Delta t)$, which a priori differs from the product of the two equilibrium probabilities. This becomes evident in the case $\Delta t = 0$, where $\underline{A}^1 = \underline{A}^2$, and thus $P(\underline{A}^1, \underline{A}^2 | \Delta t = 0) = P(\underline{A}^1)\, \delta_{\underline{A}^1, \underline{A}^2}$ with δ being the multidimensional Kronecker symbol. This extreme situation can be observed in protein families, where, e.g., protein sequences of different strains of the same species differ at most by a few mutations. Note that nevertheless each single sequence may be in equilibrium: $\sum_{\underline{A}^2} P(\underline{A}^1, \underline{A}^2 | \Delta t) = P(\underline{A}^1)$ for all Δt, and similarly for $\underline{A}^2$.

The statistical dependence between homologous proteins poses an important problem to the inference of our statistical model $P(\underline{A})$. The likelihood of the coupling and field parameters given the MSA $\{\underline{A}^m\}_{m=1...M}$ and the phylogenetic tree $\mathcal{T}$

$$\mathcal{L}(\{J_{ij}(A, B), h_i(A)\} \,|\, \{\underline{A}^m\}, \mathcal{T}) \ne \mathcal{L}_{i.i.d.}(\{J_{ij}(A, B), h_i(A)\} \,|\, \{\underline{A}^m\}) \tag{3}$$

does not factorize into a product of single-sequence probabilities $P(\underline{A}^m)$. Using the factorized expression (2) as an approximation leads to biased statistics, as groups of closely related organisms in the family lead to an over-representation of certain regions of sequence space. Two consequences are illustrated in Figure 1: if we do not consider the tree $\mathcal{T}$, columns 3 (red/blue) and 6 (green/orange) seem to have an equivalent statistics and equal single-site entropies. However, observing the tree, we see that column 3 can be explained by a single mutation in one of the early branches of the tree, whereas column 6 requires at least two mutations in more recent branches. The same amplification of mutations in subtrees may also lead to spurious correlations in the amino acid usage of column pairs. The amino acid usage of columns 3 and 8 (both red/blue) may be explained by a single mutation per site, but suggests a correlation in their joint amino acid usage. It has been recently shown [14] that the phylogenetic bias changes the spectral properties of the correlation matrix. A power-law tail of large eigenvalues emerges from the hierarchical structure the phylogenetic tree, in difference to the Marchenkov–Pastur distribution, which would be present in data lacking both phylogenetic and functional correlations.

Direct inference of a DCA model $P(\underline{A})$, by maximizing the factorized approximation of the likelihood, thus leads to the existence of field and couplings parameters that attempt to model the full biased statistics. As a result, the parameters of the DCA model cannot be expected to accurately represent functional constraints acting on the protein, even if all single sequences were individually distributed according to $P(\underline{A})$.

Usual implementations of DCA [7,8] use the so-called reweighting scheme to account for phylogeny: sequences with more than 80% identity are downweighted, counting for one observation in total. In the $\Delta t = 0$ case, this has the correct effect of considering $\underline{A}^1$ and $\underline{A}^2$ as a single observation. However, in the general setting, this is only a crude correction for the biases, which are generated by the hierarchical sequence organization on the phylogenetic tree.

Here, we aim at designing a more principled method of taking phylogenetic effects explicitly into account. This is done in Section 2, where an approximate but computationally feasible correction of phylogenetic biases is proposed. Section 2.4 discusses how the resulting corrected one- and two-site statistics can be translated into a corrected DCA model. Section 3 shows, first, results on artificial but well-controlled data, which show that our approach is able to correct the statistics of the data, and in turn to improve Potts model inference. Results on real protein data are also shown in this section. The work is concluded with a Discussion in Section 4.

2. Methods

Quantitatively, the evolutionary process can be defined by its propagator $P(\underline{A}^2|\underline{A}^1, \Delta t)$: the probability of observing sequence $\underline{A}^2$ knowing that it has sequence $\underline{A}^1$ as an ancestor at a time Δt in the past. For the evolutionary process to be stationary, the propagator should satisfy the condition

$$\sum_{\underline{A}^1} P(\underline{A}^2 \mid \underline{A}^1, \Delta t) P(\underline{A}^1) = P(\underline{A}^2) \, . \tag{4}$$

The equilibrium distribution of sequences can be recovered by taking $\Delta t \to \infty$, making sequence $\underline{A}^2$ independent from $\underline{A}^1$. Knowledge of the propagator and the phylogenetic tree would allow us to calculate the likelihood Equation (3) using Felsenstein's pruning algorithm [15]. Note that this model of evolutionary dynamics can easily take into account point mutations, deletions, and insertions, but not large-scale rearrangements like intragenic recombination, which would invalidate the assumption of the existence of a phylogenetic tree. It can take into account selection when the Hamiltonian is considered to be a fitness proxy, but it cannot take into account changes in selection, which would invalidate the assumption of stationary evolution. It can take into account changes in mutation rate when times Δt are measured in terms of a molecular clock rather then in physical time.

Assume the phylogenetic gene tree $\mathcal{T}$ to be given, with nodes indexed by n (we do not consider the problem of tree inference here). Following the description of Felsenstein's pruning algorithm in [16], let $\mathcal{L}^n(\underline{A})$ be the conditional probability of observing all existing sequences that share n as an ancestor, given that the sequence of this ancestor is $\underline{A}$, but without any information on the sequences at potential intermediary nodes inside the subtree of $\mathcal{T}$ rooted in n. If n itself represents a leaf node, i.e., an existing sequence $\underline{A}^n$, we trivially have $\mathcal{L}^n(\underline{A}) = \delta_{\underline{A},\underline{A}^n}$. For any internal node of the tree, we find the recursion relation illustrated in Figure 2A:

$$\mathcal{L}^n(\underline{A}) = \prod_{m \in \mathcal{C}(n)} \left[\sum_{\underline{B}} P(\underline{B}|\underline{A}, \Delta t^m) \mathcal{L}^m(\underline{B}) \right] , \tag{5}$$

where $\mathcal{C}(n)$ collects the children nodes of n and Δt^m equals the time separating node m from its direct ancestor n. This recursion can be conducted from the leaves to the root r of the tree, with $\mathcal{L}^r(\underline{A})$ as a result. As the sequence of the root of the tree is unknown, it is necessary to sum one more time over all possibilities for this sequence. The probability of observing the sequences of the initial MSA given the tree $\mathcal{T}$ and the model parameters, or equivalently the likelihood of the parameters given the MSA and the tree, is given as

$$\mathcal{L}(\{J_{ij}(A, B), h_i(A)\} \,|\, \{\underline{A}^m\}, \mathcal{T}) = \sum_{\underline{A}} P(\underline{A}) \mathcal{L}^r(\underline{A}) , \tag{6}$$

which obviously differs from the factorized likelihood in Equation (2). Note that in this last equation we have assumed that the propagator depends on the model parameters, i.e., the couplings $J_{ij}(A, B)$ and the fields $h_i(A)$. If we would know this dependence explicitly, we might maximize the likelihood in Equation (6) to infer the equilibrium Potts model Equation (1) from data.

However, this approach suffers from two major technical problems:

- The first is that the propagator $P(\underline{A}^2|\underline{A}^1, \Delta t)$ associated to the Potts model is not known a priori. Many distinct microscopic dynamics might lead to the same equilibrium, but the exact evolutionary processes underlying correlated protein evolution are not known. Even if we would assume some dynamics, the propagator for arbitrary time differences Δt would require to sum over all possible evolutionary trajectories going from $\underline{A}^1$ to $\underline{A}^2$—but this is intractable in practice.
- The second problem is that each use of the recursion relation (5) involves the summation over all possible sequences for each child node of node n. This amounts to summing over 21^L terms each time, with L being the sequence length.

Thus, a direct application of this scheme appears impossible for systems of realistic sizes, i.e., for typical sequence lengths L = 50–500. The following sections therefore propose two approximations based on the previously described idea, intending to make the computation of the likelihood tractable.

2.1. Approximating Dynamics: Independent-Site Evolution

To reduce the complexity of the problem, we first apply an approximation commonly used in evolutionary biology and phylogeny. The independent-site approximation—also referred to as "single-site" approximation in the following—considers each column of the MSA as evolving independently from all others. In this setting, instead of considering probabilities of observing full sequences as in $\mathcal{L}^n(\underline{A})$, we focus on the distribution of amino acids in one MSA column only. The single-site equivalent of Equation (5) becomes

$$\mathcal{L}_i^n(A) = \prod_{m \in \mathcal{C}(n)} \left(\sum_{B \in \mathcal{A}} P(B|A, \Delta t^m) \mathcal{L}_i^m(B) \right) , \tag{7}$$

where $\mathcal{L}_i^n(A)$ is the probability of observing the state of column i in existing sequences that share n as an ancestor, given that the sequence of this ancestor contains $A \in \mathcal{A}$ at this position. Summations over all possible configurations of internal nodes are replaced by summations over single symbols B, resulting in a complexity of $\mathcal{O}(L \times M \times q)$ for computing the L sitewise likelihoods. As the number M

of sequences equals the number of leaves, the number of internal nodes to be summed over equals $M - 1$.

To apply this idea, a propagator is designed using the Felsenstein model of evolution [15] and assuming a constant mutation rate μ (remember that time was measured according to a molecular clock, i.e., the assumption of constant μ is quite natural). In a time interval Δt, no mutations appear, thus with probability $e^{-\mu \Delta t}$, and B remains equal to the ancestral amino acid A. With probability $(1 - e^{-\mu \Delta t})$, one or more mutations happen. In this case, the new amino acid at position i is assumed to be chosen according to its stationary distribution $P_i(B) = \omega_i(B)$. The following propagator summarizes this process,

$$P_i(B|A, \Delta t) = e^{-\mu \Delta t} \delta_{A,B} + (1 - e^{-\mu \Delta t}) \, \omega_i(B) \, . \tag{8}$$

When using this simple dynamical model and applying the recursion of Equation (7), it is possible to compute the likelihood of the observed data very efficiently.

The likelihood does not only depend on the MSA and the phylogenetic tree, but also on the value of the mutation rate μ, which in general may be unknown. Within the independent-site approximation, we can easily estimate it using the data. To this aim, we observe that the average of the Hamming distance

$$d_H(\underline{A}, \underline{B}) = \sum_{i=1}^{L} (1 - \delta_{A_i, B_i}) \tag{9}$$

between two equilibrium sequences at evolutionary time distance Δt can be easily calculated,

$$\begin{aligned}
\bar{d}_H(\Delta t) &= \sum_{i=1}^{L} \sum_{A_i, B_i \in \mathcal{A}} (1 - \delta_{A_i, B_i}) \, P_i(A_i|B_i, \Delta t) \, \omega_i(B_i) \\
&= \left(1 - e^{-\mu \Delta t}\right) \left[L - \sum_{i,A} \omega_i(A)^2 \right] \\
&= \left(1 - e^{-\mu \Delta t}\right) \bar{d}_H(\infty) \, . \tag{10}
\end{aligned}$$

Thus, it starts at Hamming distance zero for $\Delta t = 0$, and approaches exponentially a plateau value, which is given by the average Hamming distance between two independent equilibrium sequences in the independent-site model. In the sequence data, we have no direct observation of parent–child pairs of sequences. The dynamical process given by Equation (8) is, however, a stationary one satisfying detailed balance $P_i(A|B, \Delta t) \, \omega_i(B) = P_i(B|A, \Delta t) \, \omega_i(A)$. Therefore, we can take any two sequences $\underline{A}^m, \underline{A}^n$ from the sequence alignment, calculate their Hamming distance together with their time separation on the phylogenetic tree by adding all branch lengths along their connecting path, and use the result as an instance of $d_H(\Delta t)$, cf. Figure 2B. Taking all pairs of sequences from the MSA, we can bin the observed times, calculate average Hamming distances for each time bin, and fit the functional form of Equation (10) to obtain the desired value of μ, cf. Section 3 for examples.

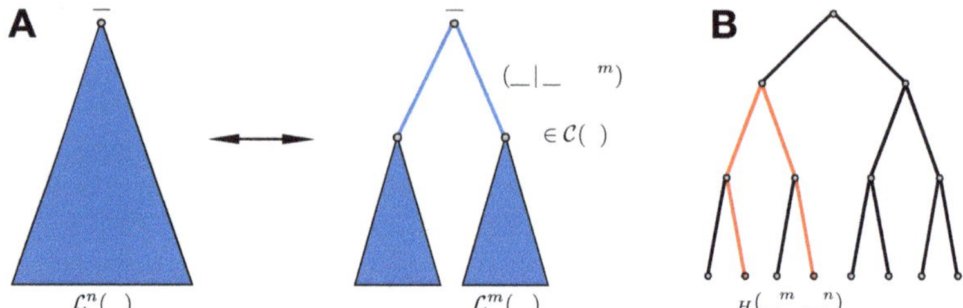

Figure 2. (**A**) Illustration of Equation (5): $\mathcal{L}^n(\underline{A})$, as represented on the left, is the probability of observing all sequences in the MSA having node n as common ancestor, given the sequence $\underline{A}$ of this ancestor. This probability can be decomposed into a product over contributions of node n's children $m \in \mathcal{C}(m)$. For each child m, we have to consider the propagator $P(\underline{B} \mid \underline{A}, \Delta t^m)$ from n to m, times the probability $\mathcal{L}^m(\underline{B})$ associated with the subtree rooted in m, and summed over all possible configurations $\underline{B}$ of m. Note that the sum over each child can be done independently; therefore, Felsenstein's algorithm runs in linear time in the number of internal nodes. (**B**) Measuring Hamming distances and time separations between sequences: thanks to the stationary dynamics of Felsentein's model, the time-dependence of the Hamming distance between a parental and a child configuration can be estimated from observed leaf configurations. To this end, for any two leaves, $\underline{A}^m$ and $\underline{A}^n$, we determine the Hamming distance $d_H(\underline{A}^m, \underline{A}^n)$ and the time separation Δt, the latter by summing the lengths of all branches on the connecting path. Time binning and averaging are used to estimate the curve $\bar{d}_H(\Delta t)$.

2.2. Approximating Dynamics: Independent-Pair Evolution

Using the independent-site approximation, one recovers the most likely single-site stationary distribution $\omega_i(A)$, given the corresponding MSA column and the topology of the evolutionary tree. Unfortunately, this method is intrinsically unable to correct for spurious correlations such as the one displayed in Figure 1. To reach that aim, we need to find a way to take two-point correlations into account. However, performing phylogenetic analysis with a model of the full sequence is intractable, as is explained at the beginning of this section.

To deal with this dilemma, we choose to use an independent-pair approximation: each pair of sites i and j is thought of as evolving independently from the others, with a propagator similar to that of Equation (8). The probability that i changes amino acid from A to C in time Δt, and j from B to D, is defined as

$$
\begin{aligned}
P_{ij}(C, D \mid A, B, \Delta t) \;=\; & e^{-2\mu\Delta t}\, \delta_{A,C}\, \delta_{B,D} \\
& + e^{-\mu\Delta t}(1 - e^{-\mu\Delta t}) \left[\omega_{j|i}(D \mid A)\, \delta_{A,C} + \omega_{i|j}(C|B)\, \delta_{B,D} \right] \\
& + (1 - e^{-\mu\Delta t})^2\, \omega_{ij}(C, D) \,,
\end{aligned}
\tag{11}
$$

where $\omega_{ij}(C, D)$ is the stationary pairwise distribution of sites i and j, and $\omega_{i|j}(C|B) = \omega_{ij}(C, B) / \sum_{C'} \omega_{ij}(C', B)$ the conditional probability of observing C in i given B in j. Note that this conditional probability is able to implement epistatic interaction between sites, in difference to the independent-site approximation. In turn, Felsenstein's recursion relation becomes

$$
\mathcal{L}^n_{ij}(A, B) = \prod_{m \in \mathcal{C}(n)} \left(\sum_{C, D \in \mathcal{A}} P(C, D \mid A, B, \Delta t^m) \mathcal{L}^m_{ij}(C, D) \right).
\tag{12}
$$

The summation over all possible configurations of two sites and the computation of the likelihood for all pairs now results in a still feasible complexity of $\mathcal{O}(L^2 \times M \times q^2)$.

Of course, a naive application of this method poses a major consistency problem: two pairs sharing one residue cannot evolve independently. As a result, the inference of the most likely pairwise statistics $\omega_{ij}(A, B)$ for each pair will give globally inconsistent results. For three pairwise distinct residues—i, j, and k—one will typically find

$$\sum_{B \in \mathcal{A}} \omega_{ij}(A, B) \neq \sum_{C \in \mathcal{A}} \omega_{ik}(A, C) , \tag{13}$$

i.e., marginal distributions for site i do not coincide when extracted from distinct pairs containing i. To settle this inconsistency, we propose a constrained optimization of the pairwise likelihoods over the probabilities ω_{ij}, subject to the constraint that its single-site marginals equal the single-site distributions obtained using the independent-site approximation scheme developed in the previous subsection (superscript "*is*"). In other words, for all i and j, the following condition is imposed,

$$\sum_{B \in \mathcal{A}} \omega_{ij}(A, B) = \omega_i^{is}(A) \quad \text{and} \quad \sum_{A \in \mathcal{A}} \omega_{ij}(A, B) = \omega_j^{is}(B) , \tag{14}$$

where $\omega_i^{is}(A)$ stands for the result of the scheme described in Section 2.1.

The hope is that by extending the phylogenetic inference beyond a sitewise description, the background pairwise statistics of the evolutionary process might be recovered, therefore improving the inference of the DCA coupling parameters.

2.3. Optimization: Maximizing the Likelihood

The independent-site or independent-pair approximations allow for a computationally efficient estimation of the likelihood. To correct empirical frequencies f for phylogenetic biases, we now need to find stationary frequencies ω maximizing the approximated likelihoods: Equation (7) (Equation (12), respectively) has to be optimized over $\omega_i(A)$ (respectively $\omega_{ij}(A, B)$). As each site i or each pair (i, j) is treated independently from the others depending on the approximation used, the optimization is conducted over either q or q^2 parameters at a time. However, the gradient of the likelihood in both approximations is intractable, and its concavity is unknown, making the use of standard gradient ascent techniques impractical.

Here, we rely on a stochastic optimization scheme, which was empirically found to be efficient in this scenario, inspired by the work in [17]. Parameter space, i.e., the $\omega_i(a)$ or the $\omega_{ij}(a, b)$, is randomly sampled by making global or local random moves: in global moves, all parameters to be optimized are simultaneously changed, while in local moves only one is changed (up to subsequent normalization). The moves are only accepted if they lead to an increased likelihood. Their magnitude is decreased throughout the optimization, starting with large displacement in parameter space and ending with small adjustments. The best parameters found are returned. This scheme is rather empirical and does not guarantee convergence. However, in testing simplified scenarios where the stationary frequencies ω are known, it was found to always lead to the correct solution.

In the case of the independent-pair approximation, $\omega_{ij}(A, B)$ needs to be optimized under the constraints defined in Equation (14). For this reason, moves proposed by the stochastic exploration of parameter space need to satisfy the constraints at all times. Here, we use a reparameterization trick inspired by the definition of direct information in [18]: tentative pair frequencies are written as

$$\omega_{ij}(A, B) = \frac{1}{z(J, \tilde{h}_i, \tilde{h}_j)} \exp \left\{ J(A, B) + \tilde{h}_i(A) + \tilde{h}_j(B) \right\} . \tag{15}$$

The optimization is then conducted over the coupling parameter J. Whenever J is changed, compensatory fields $\tilde{h}_i$ and $\tilde{h}_j$ are re-estimated to satisfy the marginalization constraints. In this way, optimization is conducted in the space of frequencies that do satisfy Equation (14).

2.4. From Corrected Frequencies to DCA Models

Our final aim is to infer DCA models of the form Equation (1), which are corrected for phylogenetic biases. In the last section, we have described an approximation scheme for correcting the single- and two-site equilibrium frequencies. These must, in a next step, be included in an inference procedure for the couplings and fields of Equation (1).

A first simple idea would be to use mean-field DCA [8], i.e., to invert the inferred covariance matrix $C_{ij}(A, B) = \omega_{ij}(A, B) - \omega_i(A)\omega_j(B)$ to obtain the coupling parameters $J_{ij}(A, B)$. However, there is a problem: even if we have constructed the $\omega_{ij}(A, B)$ carefully to obtain local coherence via fixing their single-site marginals to the $\omega_i(A)$ obtained applying the independent-site approximation, they are not globally coherent. In particular, the before-mentioned covariance matrix $C_{ij}(A, B)$ cannot be obtained as the data-covariance matrix of a sequence sample. This is easiest visible when observing the eigenvalue spectrum of the inferred C-matrix, which typically contains negative eigenvalues, while a data-covariance matrix is guaranteed to be positive semidefinite. Mean-field DCA uses positive pseudocounts for regularized inference, but this procedure would shift the negative eigenvalues of C towards larger values, and induce singularities in its in inverse.

The other popular implementation of DCA is using pseudo-likelihood maximization (plmDCA) to estimate the coupling and field parameters [19,20]. Although being more accurate than mean-field DCA, it does not use empirical single- and two-site frequencies as inputs, but the full-length sequences of the input MSA itself. To use plmDCA, we designed a way to construct an artificial MSA, which has approximately a given pairwise target statistics ω_{ij}^{target}, using a simulated annealing strategy based on the work in [21]. In a first step, we emit an MSA having the correct target profile ω_i^{target}, i.e., each column is generated independently as a sample of ω_i^{target}. In a second step, entries inside columns are permuted in a way to establish also the target correlations contained in ω_{ij}^{target}, while conserving the single-site profile ω_i^{target}: in each move t, a column i and two rows m and n are chosen at random, and an attempt to exchange A_i^m and A_i^n is made. The probability of the exchange to take place is given by the Metropolis rule:

$$P(exchange) = \min\left[1, \exp\left(-\beta||C^{t+1} - C^{target}|| + \beta||C^t - C^{target}||\right)\right], \tag{16}$$

where C^t and C^{t+1} are the covariance matrices of the current MSA before and after the exchange, and C^{target} the covariance matrix corresponding to the target frequencies. $|| \cdot ||$ stands for the Frobenius norm of matrices, and β is a formal inverse-temperature parameter. Thus, a move is more likely to be accepted if it makes the connected correlation matrix of the alignment closer to that of the target. Parameter β is initialized at a low value and slowly increased as more moves are made. In this way, when β goes to infinity, we hope to have C approaching C^{target} as much as possible (remember that our target C^{target} cannot be reached by C, as only the latter is positive semidefinite).

This procedure allows us to construct a sample approximating the corrected pairwise frequencies ω_{ij}, using the independent-pair approximation described above: the target frequencies are simply set to the ones resulting from the optimization of the likelihood, $\omega_{ij}^{target} = \omega_{ij}$. However, this is not possible when using the independent-site correction, since only the single site frequencies ω_i are corrected. In this case, we build an artificial pairwise frequency matrix defined by

$$\omega_{ij}(A, B) = f_{ij}(A, B) - f_i(A)f_j(B) + \omega_i(A)\omega_j(B), \tag{17}$$

where $f_i(A)$ is the fraction of sequences in the MSA having amino acid A in position i, and $f_{ij}(A, B)$ is the fraction of sequences having simultaneously amino acids A and B in positions i and j. The pairwise statistics defined in this way has the corrected single-site frequencies as marginals, but uncorrected connected correlations. However, a major drawback of this method is that this manner of combining different frequencies gives rise to inconsistencies, with some terms $\omega_{ij}(a, b)$ being larger than 1 or smaller than 0. It is therefore impossible for our simulated annealing procedure to construct an alignment exactly reproducing these frequencies.

Once the corrected pairwise statistics are computed following Section 2, and a corresponding MSA is built, standard plmDCA is used to infer the Potts model (1).

3. Results

3.1. Design of a Toy Model

To test the methodology, we first try our methods on a toy model. This allows us to fully control the data generation, and the true model is known. As the aim of correcting data for phylogenetic bias is ultimately to have a better DCA inference, we choose our toy model to be of the Potts form. In this manner we know that using a sufficiently large i.i.d. sample the model parameters J_{ij} and h_i can be recovered with high accuracy.

For computational efficiency, the length of the model is restricted to $L = 25$, with $q = 4$ states for its variables. Couplings and fields are drawn from a normal distribution, with couplings taking a predominantly ferromagnetic form:

$$J_{ij}^0(a,b) = s_{ij} x_{ij}^J \cdot \delta_{a,b} \quad \text{and} \quad h_i^0(a) = x_i^h(a), \tag{18}$$

where $\{x_{ij}^J\}, i,j \in \{1\ldots L\}$ and $\{x_i^h(a)\}, i \in \{1\ldots L\}, a \in \{1\ldots q\}$ are Gaussian random variables:

$$x_{ij}^J \sim \mathcal{N}(\mu_J, \sigma_J) \quad \text{and} \quad x_i^h \sim \mathcal{N}(\mu_h, \sigma_h) \tag{19}$$

with $\mu_J = 0.8, \sigma_J = 0.2, \mu_h = 0$, and $\sigma_h = 0.6$. The s_{ij} are discrete binary variables taking values in $\{0,1\}$:

$$s_{ij} = \begin{cases} 1 & \text{with probability } c/L, \\ 0 & \text{with probability } 1 - c/L. \end{cases} \tag{20}$$

To mimic the effect of structural contacts, we dilute the couplings by taking a value of $c = 3$, making the graph underlying the coupling matrix a sparse random graph [22]: each site i shares a direct coupling J_{ij} with 3 other sites j on average.

The corresponding "true" model will be called $P^0(\underline{A})$ in the following, it will constitute the ground truth, against which our inference results can be tested.

3.2. Artificial Data

To simulate the effect of phylogeny, we sample the toy model P^0 using MCMC (Markov Chain Monte Carlo) simulations on a binary tree: Each branch of the tree corresponds to an independent finite-time MCMC run. For a branch of length Δt, a number of "mutations" is drawn from a Poisson distribution with mean $\mu L \Delta t$, with μ being the mutation rate per site and time unit. For each of these mutations, a site i is chosen at random and its new state is drawn from the local conditional probability $P^0(A_i|A_1, \ldots, A_{i-1}, A_{i+1}, \ldots, A_L)$ in a Gibbs-sampling manner.

To generate an MSA, first, a root configuration is drawn from P^0, duplicated onto the two outgoing branches, and the described finite-time MCMC runs are performed. This process is iterated, taking the two resulting configurations as new roots, thus growing the tree. For K iterations, the resulting tree will consequently have 2^K leaves, whose Potts configurations are reported as artificial MSA.

This scheme guarantees that the number of mutational events will correspond to dynamical models in Equations (8) and (11). However, the way residues are re-drawn after a mutation depends on the full current sequence through distribution P^0, unlike the simplifying assumptions of the propagators.

For simplicity reasons, $\mu\,L\,\Delta t$ is set to be identical for all branches of the tree, taking values 3, 5 or ∞ (i.e., $\mu\Delta t \gg 1$), resulting in respectively strong, weak and absent phylogenetic effects. In the following, the samples corresponding to finite values of Δt will be referred to as biased samples, whereas the one corresponding to $\Delta t \to \infty$ will be referred to as a fair or i.i.d. sample. 12 duplication events are performed, resulting in a tree of $2^{12} = 4096$ leaves and $2^{12} - 1$ internal nodes. Finally, so as not to depend on the particular choice of the root configuration, 30 repetitions of the sampling process are performed for each Δt.

To keep the main text concise, only results concerning the $\mu\,L\,\Delta t = 3$ are shown. This represents the hardest case, as phylogeny effects are more pronounced for short branch lengths. Results for $\mu\,L\,\Delta t = 5$ are shown in the supplementary material in the form of figures.

Note that, for a model without couplings, the data generating process would correspond exactly to the dynamics described in Section 2.1. For a coupled model, however, the real μ may differ from the one to be used to fit our independent-site or -pair models, due to a slowing down of the MCMC dynamics. We, therefore, use the strategy described in Section 2.1: For each sequence pair, the Hamming distance and the evolutionary time separation are calculated. Times are binned (in the simplified data generation times are actually discrete), and average Hamming distances are computed. The resulting data are fitted against the theoretical result in Equation (10) to obtain the effective mutational parameter to be used in the phylogenetic inference. Results for $\mu\,L\,\Delta t = 3$ are shown in Supplementary Figure S1, choosing $\Delta t = 0.3$ without loss of generality.

3.3. Phylogenetic Inference Corrects the One- and Two-Point Statistics

To assess the quality of the phylogenetic correction, we first compare single-site and pairwise statistics before and after our inference to the same observables measured in an i.i.d. sample drawn from P^0.

In the case of the independent-site approximation, the single-site statistics are corrected. Observables measured in the biased sample, i.e., the sample coming from the leaves of the tree, without correction, referred to as the "tree" sample, will be denoted as f_i^t. After phylogenetic correction, we call the single-site frequencies f_i^{inf}. The statistics of the i.i.d. sample is f_i^0, obviously without any correction applied.

As demonstrated in Figure 3, the inference clearly improves the estimation of single-site frequencies over naive counting in the biased sample. Pearson correlations between f_i^{inf} and f_i^0 are significantly higher than between f_i^t and f_i^0, being larger than 0.75 in 27 out of 30 repetitions. This contrasts with the remarkably low correlations of 0.4 that can be achieved for some realizations of the tree if no correction is performed. Similarly, the slope of a linear regression of f_i^{inf} against f_i^0 tends to be much closer to 1 in most cases, also showing lower variation from repetition to repetition.

A similar comparison is made for pairwise frequencies in the case of the independent-pair approximation. We now compare f_{ij}^t and f_{ij}^{inf} to their counterpart from the i.i.d. sample f_{ij}^0. The two top panels of Figure 4 once again show an improvement resulting from the phylogenetic inference, as pairwise statistics are closer to match f_{ij}^0 after it is performed.

However, one has to keep in mind that some of this improvement is due to the single-site correction. Indeed, in the independent-pair approximation, marginals of the pairwise frequencies are constrained to match the corrected single site frequencies f_i^{inf}. To evaluate the intrinsic quality of the pairwise method, we focus on the connected correlations $c_{ij} = f_{ij} - f_i f_j$, thus removing the influence of the single-site correction. Bottom panels of Figure 4 demonstrate that even this intrinsically pairwise quantity is recovered with higher accuracy after inference, even if to a somewhat lesser extent than for the frequencies. Even our very crude approximation—considering every pair as evolving independently—can correct some of the statistical bias due to phylogeny, improving over naive counting in the MSA.

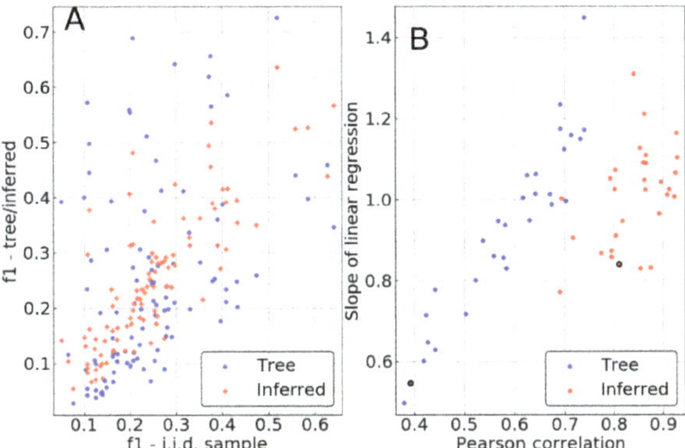

Figure 3. Result of the single-site phylogenetic inference for $\mu\, L\, \Delta t = 3$. (**A**) Single-site statistics of a sample of P^0 coming from a tree, before ("Tree"), and after ("Inferred") phylogenetic inference, against "true" single site statistics coming from the fair i.i.d. sample. (**B**) Slope of the linear regression and Pearson correlation corresponding to the plot in panel (A), for the 30 repetitions of the experiment. The black-circled points correspond to the sample displayed in panel (A).

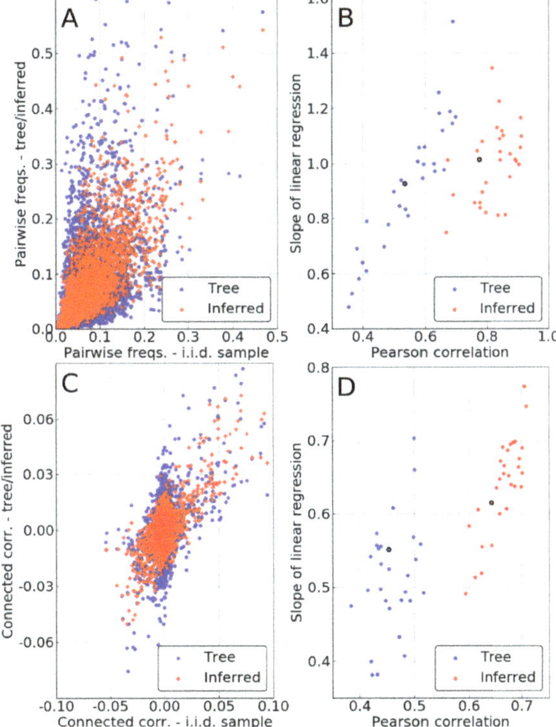

Figure 4. Result of the pairwise phylogenetic inference for $\mu\, L\, \Delta t = 3$. (**A**) Pairwise frequencies $f_{ij}(a, b)$ of a sample of P^0 coming from a tree, before ("Tree"), and after ("Inferred") the phylogenetic inference, against "true" pairwise frequencies coming from the fair sample. (**B**) Slope of the linear regression and Pearson correlation corresponding to the plot in panel (A), for the 30 repetitions of the experiment. The black-circled points correspond to the repetition displayed in panel (A). (**C**) Same as panel (A) for connected correlations $c_{ij} = f_{ij} - f_i f_j$. (**D**) Same as panel (B) for connected correlations.

3.4. DCA Parameters are Recovered with Increased Accuracy

We infer DCA models based both on the uncorrected and the corrected frequencies f_{ij}^{t} and f_{ij}^{inf} using the methodology described in Section 2.4. To evaluate both of our approximations, we infer the DCA model in the case of the single-site correction and the independent-pair correction.

In the top panel of Figure 5, inferred parameters are compared to the true ones J^0 and h^0 using Pearson correlation as a measure. Both methods—single sites and independent pairs, labeled as pairwise in the figures—lead to a significant improvement in the inference of fields. However, the inference of couplings is deteriorated when using only the single site correction, whereas it is improved in the pairwise case. This may be due to the inconsistencies appearing when combining correlations from the biased sample with corrected single site frequencies, as is explained in Section 2.4. Indeed, such inconsistencies (frequencies larger than 1 or smaller than 0) were observed for all of the 30 repetitions.

To understand if an inferred DCA model $\hat{P}$ is a good fit to the true distribution, we compute its symmetrized Kullback–Leibler (KL) divergence to the data-generating model P^0:

$$D_{KL}(\hat{P}||P^0) + D_{KL}(P^0||\hat{P}) = \langle \mathcal{H}_{\hat{P}} - \mathcal{H}_{P^0} \rangle_{P^0} + \langle \mathcal{H}_{P^0} - \mathcal{H}_{\hat{P}} \rangle_{\hat{P}} , \qquad (21)$$

where $\mathcal{H}_P$ indicates the Hamiltonian (or, up to an additive and sequence-independent term, log-probability) of a statistical model P, and $\langle \cdot \rangle_P$ the average over P. Although the standard D_{KL} depends on the intractable calculation of the partition function of one of the distributions, its symmetrized version can be easily estimated by MCMC sampling from the average energies of the two models, evaluated on samples of each model. It is a reasonable information theoretic distance measure for distributions, as it is zero if and only if the two distributions coincide, and positive otherwise. Figure 5B shows a histogram of this quantity for the 30 repetitions of the sampling process. A clear ranking between methods appears, with the inference based on the biased sample being the worst. Both phylogenetic corrections result in a model that is closer to P^0, with anadvantage for the pairwise method. Surprisingly, the decrease in inference quality of the couplings when using the single-site correction does not appear to have a strong influence on Kullback–Leibler divergence, as there is a very large drop of this quantity between a biased sample or a single-site correction based DCA.

Note also that the imperfect nature of our approximation scheme becomes visible in the figure: the KL divergence of the model inferred from an i.i.d. sample can be seen as a lower bound for what can beobtained with a finite sample. It is substantially smaller than even the pairwise correction using the same sample size.

Another important test of the model quality, in particular for protein systems, is the "contact prediction": strong couplings between pairs of sites are expected to correspond to the sparse graphical structure of the model P^0 used for data generation. To this end, couplings are ordered with respect to their coupling strength (measured by the Frobenius norm of the coupling matrix in so-called zero-sum gauge, cf. [20]); the positive predictive value (PPV) is the fraction of true predictions (nodes connected by a link in the ground truth) in between the N first predictions. It is plotted as function of N in Figure 5C. The inference based on the i.i.d. sample is perfect in this case, ranking couplings on true links before those being not adjacent in the ground truth. The inference based on the biased "tree" sample is performing slightly worse, and it is partially corrected by the pairwise correction. On the contrary, as can already be expected from Figure 5A, the single-site correction deteriorates the reconstruction of the interaction graph.

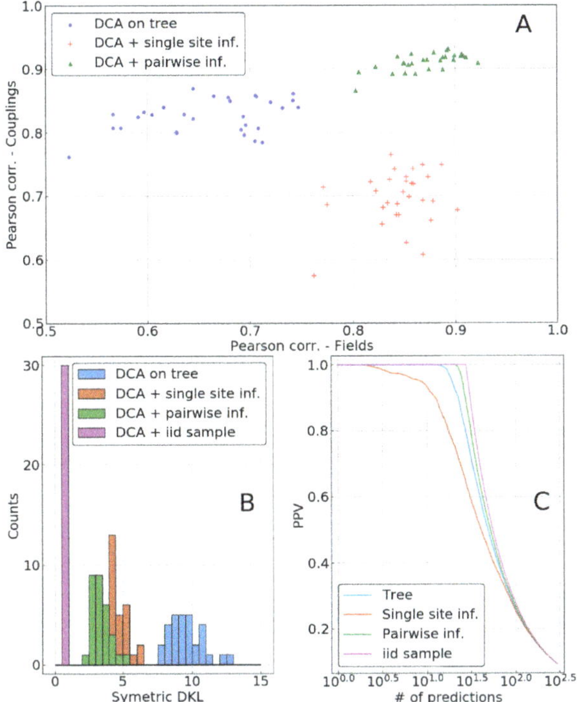

Figure 5. Direct coupling analysis (DCA) models inferred after single-site or pairwise phylogenetic correction for $\mu L \Delta t = 3$. (**A**) Pearson correlation between parameters of inferred and of true DCA models. y-axis: couplings J_{ij}; x-axis: fields h_i. One point corresponds to one repetition of the MCMC process on the tree, i.e., to one sample. (**B**) Histogram of the symmetrized Kullback–Leibler divergences between inferred and true models for all samples. (**C**) Positive predictive value for predicting non-zero couplings (i.e., "contacts") using inferred DCA models. DCA inferred on the i.i.d. sample performs perfectly in this case.

3.5. Improvement in the Prediction of Single Mutant's Energies

One of the most promising application of DCA-like methods is their ability to infer the effect of mutations in proteins from the MSA of diverged homologs [23–27]. Here, we investigate the potential of our phylogenetic correction to enhance the accuracy of these predictions. To recreate this setting in our toy model, we consider single-site "mutants" of "wild type" artificial sequences. Wild types can be taken either in the phylogenetically biased sample, as would be the case in standard DCA, or in the i.i.d. sample, i.e., without phylogenetic correlation to the sequences in the MSA. For any wild type sequence $\underline{A} = (A_1, ..., A_L)$, the $L \times (q-1)$ single mutants (i.e., single-spin flips) are denoted by $\underline{A}_{(i,\alpha)}$, with $i \in \{1, ..., L\}$, $\alpha \in \mathcal{A} \setminus A_i$. For each of these, the effect of the mutation is defined by the energy difference between wild type $\underline{A}$ and mutant $\underline{A}_{(i,\alpha)}$:

$$\Delta \mathcal{H}_{i\alpha} = \mathcal{H}(\underline{A}_{(i,\alpha)}) - \mathcal{H}(\underline{A}) . \tag{22}$$

$\mathcal{H}$ can be either the true Hamiltonian $\mathcal{H}^0$ of the generative model P^0, then defining the true mutational effect, or an inferred one, corresponding to the predicted mutational effect.

To evaluate the influence of both the phylogenetic correction and the DCA methodology on the quality of predictions, we choose to also infer a profile model as a comparison point. Profile models have vanishing couplings and reproduce the single site statistics $f_i(A) \sim e^{h_i(A)}$ using local fields only, different sites are independent. They have been used with success for predicting mutational effects in proteins based on the conservation profile of the MSA [28,29], and they are the asymptotic stationary distributions of the independent-site evolution model of Section 2.1.

We first focus on the single-site phylogenetic corrections. Given a model (profile/DCA), a statistics (tree/corrected), and a specific wild type sequence $\underline{A}$, we compute the Pearson correlation between the predicted energy shifts $\{\Delta H_{(i,\alpha)} \mid i \in \{1,...,L\}, \alpha \in \mathcal{A} \setminus A_i\}$ and the true ones, $\{\Delta H^0_{(i,\alpha)}\}$. This is done for all sequences either in the biased or the i.i.d. sample, and resulting correlations are averaged over each sample. The resulting value represents thus the quality of predictions of the energies of single mutants with wild types in a given sample.

As is shown in Figure 6, when the reference sequence is taken in the biased sample, all methods seem to perform equally well, apart from the profile model inferred on the biased frequencies. In particular, applying the DCA methodology and thus attempting to fit correlations or using a simple profile model on corrected data seems to result in the same improvement.

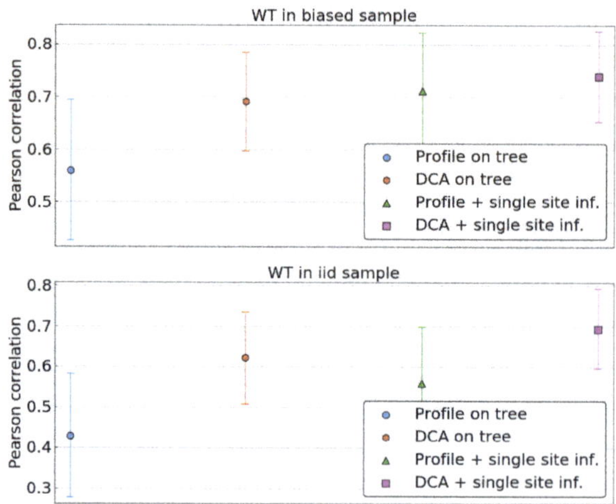

Figure 6. Pearson correlation in predicting energies of single mutants averaged over sets of reference sequences for $\mu L \Delta t = 3$. In the top panel, reference sequences are taken in the biased sample, i.e., among the leaves of the phylogenetic tree. In the bottom panel, reference sequences are taken in a fair sample of P^0. Predictions are made using four models: a profile model and a Potts model trained on the uncorrected biased sample, respectively ("Profile on tree" and "DCA on tree", respectively), and using the corrected single site frequencies ("Profile + single site inf." and "DCA + single site inf.", respectively). Error bars indicate the standard deviation across the 30 repetitions of the tree sampling process.

The picture changes when the reference sequence is taken in a fair sample, i.e., when it is independent from the sample used for model inference. In this case, the performance of both DCA on uncorrected data and of the profile models drop significantly, whereas DCA inferred on corrected frequencies remains accurate. To investigate this further, we compute the average Pearson correlation as a function of the Hamming distance of the wild type to the closest sequence in the biased sample. Supplementary Figure S2 shows that while the performance of the uncorrected DCA and the profile models declines rapidly when using a reference sequence far away from the biased sample, the corrected DCA has a more stable performance before large Hamming distances are reached.

As the combination of DCA and of the single site phylogenetic correction outperforms profile models or a naive DCA approach, we now consider inferring the Potts model based on the corrected pairwise frequencies. The same scoring as above is used, using all single mutants for wild type sequences in both samples and computing the average Pearson correlation across wild types. Figure 7 compares the predictions of the DCA models using the tree levels of phylogenetic correction: none, sitewise and pairwise. The latter leads to a significant improvement in accuracy of predictions, outperforming the two other methods. This stands both in the case of a wild type belonging to the biased sample or to the fair sample.

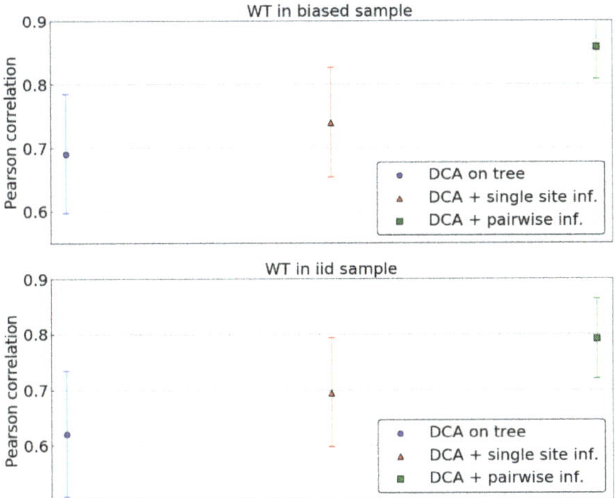

Figure 7. Pearson correlation in predicting energies of single mutants for $\mu\,L\,\Delta t = 3$ averaged over sets of reference sequences. In the top panel, reference sequences are taken in the biased sample, i.e., among the leaves of the phylogenetic tree. In the bottom panel, reference sequences are taken in a fair sample of P^0. Predictions are made using a DCA model inferred either directly on biased data, either using corrected single site frequencies, either using corrected pairwise frequencies. Error bars indicate the standard deviation across the 30 repetitions of the tree sampling process.

Again, we investigate the dependence of those predictions on the distance of the wild type to the closest sequence in the biased sample. The largest increase in Pearson correlation resulting from the pairwise phylogenetic inference once again happens for sequences that are far from the biased sample (Figure S3). Removing part of the phylogenetic bias seems to have a stronger influence when considering the energy landscape around sequences that are far away from the leaves of the phylogenetic tree. When using those leaves as a sample without accounting for their non-independence, the resulting model seems not to learn much about the energy landscape far away from those points. However, correcting for non-independence, even in a rather crude way, leads to a much better inference in this regard.

3.6. Results on Protein Data

The main application of DCA-like methods so far has been their ability to predict contacts in the three-dimensional protein structure. Strong couplings between two sites in the Potts model are a good indication of the corresponding amino acids being in contact in the protein fold. As, in the case of artificial data, couplings are inferred more accurately when frequencies are corrected for phylogeny (Figure 5), it is natural to ask whether this translates to improved contact predictions for actual protein data.

To assess the performance of our correction scheme on actual protein data, we evaluated the PPV of DCA contact predictions on five protein families(cf. Supplementary Material S1 for details). Those families were chosen from the families used in [8] on the basis of having short enough sequences for our pairwise phylogenetic correction to be tractable in reasonable time, and to have potentially stronger phylogenetic correlations than current Pfam data, which are based on representative proteomes, i.e., which have undergone already some phylogeny-based sequence pruning. In contrast to the artificial data, the phylogenetic tree is not a priori known, and we have applied FastTree [30,31] for each family for tree inference. Next, for each family, three DCA models were inferred: a "naive" model based on completely uncorrected statistics; a model based on frequencies corrected by the reweighting scheme, which is the one used in common DCA implementations; and a model based on frequencies corrected by our pairwise phylogenetic inference scheme. Contact prediction was done using the standard procedure of plmDCA [20].

Figure 8 shows representative results for two of the five families. In the case of PF00013, our phylogenetic correction clearly performs worse than both reweighted and uncorrected DCA for the first 100 predictions. Note that the reweighting method does not lead to any improvement either, suggesting that the phylogenetic bias may be weak for this family, and the potential benefit of the correction is overcome by problems due to the approximations used. The picture changes for PF00046, where both correction methods—reweighting and phylogenetic inference—improve significantly over the uncorrected DCA model. Reweighting outperforms our method for the prediction corresponding to the strongest coupling, having a fraction of true predictions of ∼0.7 versus ∼0.5 for the first ten predictions. However, for a large number of predictions, the phylogenetically informed DCA model tends to have an enriched fraction of contacts among its couplings when compared to the reweighted model. This observation fits well with results on artificial data, showing an overall increase in the accuracy of inferred couplings. However, as applications of DCA usually rely on the very strong couplings only, this long-term increase in accuracy remains of limited practical interest.

Results for three other families can be seen in Figure S4. Over all the five investigated protein families, our phylogenetic correction only shows improvement with respect to an uncorrected model for two of them: PF00046 and PF00111. In both cases, it is outperformed by reweighting in the first predictions.

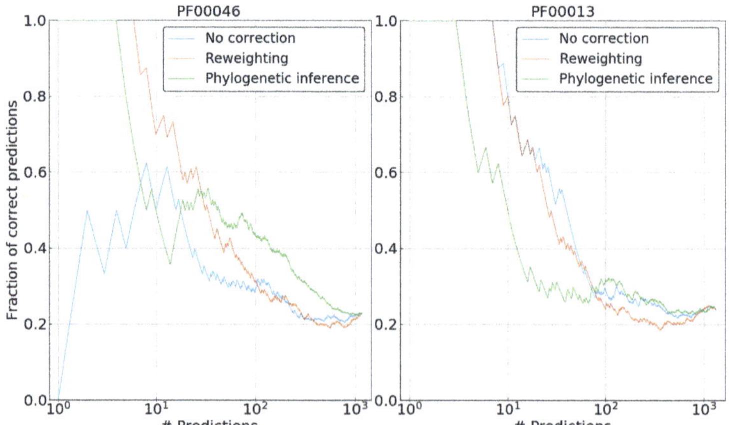

Figure 8. Positive predictive value for predicting contacts in representative structures for two protein families PF00013 and PF00046. The blue lines indicate a naive DCA method without any correction for phylogeny. The orange lines show results for the sequence reweighting scheme. The green lines show results after our phylogenetic inference scheme.

4. Discussion

In this paper, we propose a principled way to correct for phylogenetic effects in the inference of Potts models from sequence data. Although the standard technique to account for these effects in coevolutionary analyses relies on an empirical reweighting of sequences, our method aims at doing so using the phylogenetic tree as well as an evolutionary model. The global nature of Potts models implies that the evolutionary model used should depend on the full sequence. However, such a global approach is intractable in the case of discrete variables such as amino acids. To overcome this problem, we proposed two subsequent levels of approximation: the first one relying on sites evolving independently as in standard models of sequence evolution; the second one describing pairs of sites, which display internal correlations but evolve independently from the rest of the sequence.

We show that our phylogenetic correction method combined with these approximations is efficient in the case of artificial data. When data are generated by a known Potts model using a sensible but simple evolutionary process on a known tree, our method is able to efficiently correct single-site and pairwise statistics, including connected correlations as intrinsically pairwise quantities. This, in turn, results in an improved inference of the Potts model in all tested aspects: individual coupling and field parameters are more accurate, the inferred Potts probability distribution is closer to the real one, contact prediction is more precise, and prediction of local energy changes from mutations is improved.

In the case of actual protein families however, results are at best mitigated. For only two of the five investigated families (PF00046 and PF00111), our method does improve the accuracy of contact predictions with respect to uncorrected data, whereas it has a negative effect on these predictions for two of the other families (PF00013 and PF00014). Furthermore, even in the positive cases, it is still outperformed by the simpler empirical method of reweighting sequences according to simple sequence-similarity measures.

In this regard, it is important to note two things: The first is that for the two families for which our method fails, the reweighting technique leads to very marginal improvements in terms of contact prediction. This seems to indicate that our method does perform reasonably well only in the case of strong phylogenetic biases. It also suggests that phylogeny does not affect contact prediction to a noticeable degree in some families. The second is that in the cases of protein families for which our method does provide an improvement, it outperforms reweighting in the "long run", e.g., for more than $\sim$100 predictions for PF00046. This may mean that phylogeny has a strong effect on weaker DCA couplings that reweighting fails to correct. Even though these are not necessarily relevant for contact prediction, they impact the accuracy of the model in other aspects, such as predicting mutational effects or generating new sequences. If one wants to use DCA as a sequence model rather than simply a contact-prediction tool, it becomes all the more important to correct for phylogeny if it has a global influence on all parameters of the model. If this is the case, it is arguable that principled methods such as ours would be more appropriate than uninformed methods such as reweighting at correcting subtle effects of phylogeny.

Different reasons can be invoked for the mitigated results on protein families. One is that our method relies on the exact knowledge of the phylogenetic tree, depending both on its topology and on branch lengths. This knowledge is of course not available for proteins, where we rely on inference software to find a tree. Inaccuracies in this tree inevitably affect our method in a negative way. Another possible problem is the stationary and Markovian nature of our evolutionary model, which may not be true in the case of protein evolution. Over evolutionary time scales, variable environments lead to changing selective pressures, population sizes and mutation rates, which are currently not accounted for by our model. However, we expect that the major problem lies in the nature of the approximations we had to resort to. The first one, independent sites, is in contradiction with the global nature of the Potts model we try to infer. The second—independent pairs—allows for the correction of pairwise statistics, but suffers from obvious consistency problems since overlapping pairs of sites cannot be considered independent. Note that this has an important consequence, when we go to protein families with longer sequences: whereas phylogenetic tree inference becomes more accurate for longer sequences, and the independent pair approximation requires $\mathcal{O}(L^2)$ inferences for all pairs of residue positions, with L being the sequence length. As a consequence, the before-mentioned inconsistencies are expected to grow drastically with sequence length.

The necessity for these approximations comes from two characteristics of the class of models we are using: their global nature, in the sense that they[37] give probabilities to sequences in a nonfactorized way, and the discrete nature of the variables used (amino acids). By rendering certain calculations intractable, such as tracing over all possible states of internal nodes in the tree; these two characteristics make the use of approximations unavoidable. In this article, approximations attempt to circumvent the global nature of the Potts model by factorizing probabilities in different ways, namely, sitewise and pairwise. However, an interesting different class of approximations would be to forget about the discrete nature of amino acids and model them by continuous variables instead. This would transform the Potts model into a Gaussian distribution, making the design of a global propagator tractable. Note that on similar grounds gaussDCA [32], an analytically solvable Gaussian version of DCA, was developed a few years back, and it was found to perform similar to other DCA techniques in contact prediction.

Another interesting alternative might be built upon the observation made in [14]: phylogenetic correlations between the sequences of the training MSA lead to a fat tail of large eigenvalues of the covariance matrix, i.e., of the empirically observed statistics reproduced by DCA models. Furthermore, it was argued in [33] that the corresponding eigenvectors are extended over many positions and amino acids, thereby giving rise to many small couplings. The contact prediction was found to be more closely related to small eigenvalues of the covariance matrix, with localized eigenvectors giving rise to large localized couplings. However, while phylogenetic correlations between sequences are sufficient to generate extended eigenvectors with large eigenvalues, the latter may also result from slightly different functionalities of subfamilies of the studied MSA, i.e., they may contain biologically sensible information, cf. [34,35]. Disentangling the two—sequence clustering by phylogeny and by subfunctionalization—seems a nontrivial task.

As DCA-like pairwise models are increasingly used in sequence analysis, and as their ability to accurately model sequence variability in protein families gets more established, the need to infer parameters more accurately and without bias increases. For this reason, correcting for phylogeny in a controlled and principled way is essential. Whether this can be achieved using techniques similar to the one presented in this paper, or using different types of approximations as the two mentioned in the last to paragraphs, or totally different techniques, remains a widely open and challenging question.

Supplementary Materials: The following are available at http://www.mdpi.com/1099-4300/21/11/1090/s1: Supplementary Material S1, containing Figures S1–S11, Table S1.

Author Contributions: Conceptualization, M.W.; methodology, E.R.H., P.B.-C., and M.W.; numerical implementation and analysis, E.R.H. and P.B.-C.; investigation, E.R.H., P.B.-C., and M.W.; writing—original draft preparation, E.R.H., P.B.-C., and M.W.; writing—review and editing, P.B.-C. and M.W.; supervision, M.W.; funding acquisition, M.W.

Funding: This work was funded by the EU H2020 Research and Innovation Programme MSCA-RISE-2016 under Grant Agreement No. 734439 InferNet.

Acknowledgments: We acknowledge helpful discussions with Alejandro Lage and Roberto Mulet.

Conflicts of Interest: The authors declare no conflicts of interest.

Abbreviations

The following abbreviations are used in this manuscript:

DCA	Direct Coupling Analysis
MCMC	Markov Chain Monte Carlo
MSA	Multiple Sequence Alignment
PPV	Positive Predictive Value

References

1. Consortium, U. UniProt: A worldwide hub of protein knowledge. *Nucleic Acids Res.* **2018**, *47*, D506–D515. [CrossRef]
2. Reddy, T.B.; Thomas, A.D.; Stamatis, D.; Bertsch, J.; Isbandi, M.; Jansson, J.; Mallajosyula, J.; Pagani, I.; Lobos, E.A.; Kyrpides, N.C. The Genomes OnLine Database (GOLD) v. 5: A metadata management system based on a four level (meta) genome project classification. *Nucleic Acids Res.* **2014**, *43*, D1099–D1106. [CrossRef]

3. El-Gebali, S.; Mistry, J.; Bateman, A.; Eddy, S.R.; Luciani, A.; Potter, S.C.; Qureshi, M.; Richardson, L.J.; Salazar, G.A.; Smart, A.; et al. The Pfam protein families database in 2019. *Nucleic Acids Res.* **2018**, *47*, D427–D432. [CrossRef]

4. Eddy, S.R. Profile hidden Markov models. *Bioinform. (Oxf. Engl.)* **1998**, *14*, 755–763. [CrossRef] [PubMed]

5. Durbin, R.; Eddy, S.R.; Krogh, A.; Mitchison, G. *Biological Sequence Analysis: Probabilistic Models of Proteins and Nucleic Acids*; Cambridge University Press: Cambridge, UK, 1998.

6. De Juan, D.; Pazos, F.; Valencia, A. Emerging methods in protein co-evolution. *Nat. Rev. Genet.* **2013**, *14*, 249. [CrossRef] [PubMed]

7. Cocco, S.; Feinauer, C.; Figliuzzi, M.; Monasson, R.; Weigt, M. Inverse statistical physics of protein sequences: A key issues review. *Rep. Prog. Phys.* **2018**, *81*, 032601. [CrossRef] [PubMed]

8. Morcos, F.; Pagnani, A.; Lunt, B.; Bertolino, A.; Marks, D.S.; Sander, C.; Zecchina, R.; Onuchic, J.N.; Hwa, T.; Weigt, M. Direct-coupling analysis of residue coevolution captures native contacts across many protein families. *Proc. Natl. Acad. Sci. USA* **2011**, *108*, E1293–E1301. [CrossRef] [PubMed]

9. Nguyen, H.C.; Zecchina, R.; Berg, J. Inverse statistical problems: From the inverse Ising problem to data science. *Adv. Phys.* **2017**, *66*, 197–261. [CrossRef]

10. Marks, D.S.; Hopf, T.A.; Sander, C. Protein structure prediction from sequence variation. *Nat. Biotechnol.* **2012**, *30*, 1072. [CrossRef] [PubMed]

11. Ovchinnikov, S.; Park, H.; Varghese, N.; Huang, P.S.; Pavlopoulos, G.A.; Kim, D.E.; Kamisetty, H.; Kyrpides, N.C.; Baker, D. Protein structure determination using metagenome sequence data. *Science* **2017**, *355*, 294–298. [CrossRef]

12. Levy, R.M.; Haldane, A.; Flynn, W.F. Potts Hamiltonian models of protein co-variation, free energy landscapes, and evolutionary fitness. *Curr. Opin. Struct. Biol.* **2017**, *43*, 55–62. [CrossRef] [PubMed]

13. Felsenstein, J. *Inferring Phylogenies*; Sinauer Associates Sunderland: Sunderland, MA, USA, 2004; Volume 2.

14. Qin, C.; Colwell, L.J. Power Law Tails in Phylogenetic Systems. *Proc. Natl. Acad. Sci. USA* **2018**, *115*, 690–695. [CrossRef] [PubMed]

15. Felsenstein, J. Evolutionary trees from DNA sequences: A maximum likelihood approach. *J. Mol. Evol.* **1981**, *17*, 368–376. [CrossRef]

16. van Nimwegen, E. Finding regulatory elements and regulatory motifs: A general probabilistic framework. *BMC Bioinform.* **2007**, *8*, S4. [CrossRef]

17. Delgoda, R.; Pulfer, J.D. A guided Monte Carlo search algorithm for global optimization of multidimensional functions. *J. Chem. Inf. Comput. Sci.* **1998**, *38*, 1087–1095. [CrossRef]

18. Weigt, M.; White, R.A.; Szurmant, H.; Hoch, J.A.; Hwa, T. Identification of direct residue contacts in protein–protein interaction by message passing. *Proc. Natl. Acad. Sci. USA* **2009**, *106*, 67–72. [CrossRef]

19. Balakrishnan, S.; Kamisetty, H.; Carbonell, J.G.; Lee, S.I.; Langmead, C.J. Learning generative models for protein fold families. *Proteins Struct. Funct. Bioinform.* **2011**, *79*, 1061–1078. [CrossRef]

20. Ekeberg, M.; Lövkvist, C.; Lan, Y.; Weigt, M.; Aurell, E. Improved contact prediction in proteins: Using pseudolikelihoods to infer Potts models. *Phys. Rev. E* **2013**, *87*, 012707. [CrossRef]

21. Socolich, M.; Lockless, S.W.; Russ, W.P.; Lee, H.; Gardner, K.H.; Ranganathan, R. Evolutionary information for specifying a protein fold. *Nature* **2005**, *437*, 512. [CrossRef]

22. Erdős, P.; Rényi, A. On the evolution of random graphs. *Publ. Math. Inst. Hung. Acad. Sci.* **1960**, *5*, 17–60.

23. Mann, J.K.; Barton, J.P.; Ferguson, A.L.; Omarjee, S.; Walker, B.D.; Chakraborty, A.; Ndung'u, T. The Fitness Landscape of HIV-1 Gag: Advanced Modeling Approaches and Validation of Model Predictions by In Vitro Testing. *PLoS Comput. Biol.* **2014**, *10*, e1003776. [CrossRef] [PubMed]

24. Morcos, F.; Schafer, N.P.; Cheng, R.R.; Onuchic, J.N.; Wolynes, P.G. Coevolutionary information, protein folding landscapes, and the thermodynamics of natural selection. *Proc. Natl. Acad. Sci. USA* **2014**, *111*, 12408–12413. [CrossRef] [PubMed]

25. Figliuzzi, M.; Jacquier, H.; Schug, A.; Tenaillon, O.; Weigt, M. Coevolutionary landscape inference and the context-dependence of mutations in beta-lactamase TEM-1. *Mol. Biol. Evol.* **2016**, *33*, 268–280. [CrossRef] [PubMed]

26. Hopf, T.A.; Ingraham, J.B.; Poelwijk, F.J.; Schärfe, C.P.; Springer, M.; Sander, C.; Marks, D.S. Mutation effects predicted from sequence co-variation. *Nat. Biotechnol.* **2017**, *35*, 128–135. [CrossRef]

27. Feinauer, C.; Weigt, M. Context-Aware Prediction of Pathogenicity of Missense Mutations Involved in Human Disease. *arXiv* **2017**, arXiv:1701.07246.

28. Ng, P.C.; Henikoff, S. SIFT: Predicting amino acid changes that affect protein function. *Nucleic Acids Res.* **2003**, *31*, 3812–3814. [CrossRef]
29. Adzhubei, I.A.; Schmidt, S.; Peshkin, L.; Ramensky, V.E.; Gerasimova, A.; Bork, P.; Kondrashov, A.S.; Sunyaev, S.R. A method and server for predicting damaging missense mutations. *Nat. Methods* **2010**, *7*, 248. [CrossRef]
30. Price, M.N.; Dehal, P.S.; Arkin, A.P. FastTree: Computing large minimum evolution trees with profiles instead of a distance matrix. *Mol. Biol. Evol.* **2009**, *26*, 1641–1650. [CrossRef]
31. Price, M.N.; Dehal, P.S.; Arkin, A.P. FastTree 2–approximately maximum-likelihood trees for large alignments. *PLoS ONE* **2010**, *5*, e9490. [CrossRef]
32. Baldassi, C.; Zamparo, M.; Feinauer, C.; Procaccini, A.; Zecchina, R.; Weigt, M.; Pagnani, A. Fast and accurate multivariate Gaussian modeling of protein families: Predicting residue contacts and protein-interaction partners. *PLoS ONE* **2014**, *9*, e92721. [CrossRef]
33. Cocco, S.; Monasson, R.; Weigt, M. From principal component to direct coupling analysis of coevolution in proteins: Low-eigenvalue modes are needed for structure prediction. *PLoS Comput. Biol.* **2013**, *9*, e1003176. [CrossRef] [PubMed]
34. Tubiana, J.; Cocco, S.; Monasson, R. Learning protein constitutive motifs from sequence data. *eLife* **2019**, *8*, e39397. [CrossRef] [PubMed]
35. Shimagaki, K.; Weigt, M. Selection of sequence motifs and generative Hopfield-Potts models for protein families. *Phys. Rev. E* **2019**, *100*, 032128. [CrossRef] [PubMed]

entropy 

Article

Coevolutionary Analysis of Protein Subfamilies by Sequence Reweighting

Duccio Malinverni [1,*] and **Alessandro Barducci** [2,*]

1   Medical Research Council (MRC) Laboratory of Molecular Biology, Cambridge CB20QH, UK
2   Centre de Biochimie Structurale (CBS), INSERM, CNRS, Université de Montpellier, 34090 Montpellier, France
*   Correspondence: ducciom@mrc-lmb.cam.ac.uk (D.M.); alessandro.barducci@inserm.fr (A.B.)

Received: 17 October 2019; Accepted: 14 November 2019; Published: 16 November 2019

Abstract: Extracting structural information from sequence co-variation has become a common computational biology practice in the recent years, mainly due to the availability of large sequence alignments of protein families. However, identifying features that are specific to sub-classes and not shared by all members of the family using sequence-based approaches has remained an elusive problem. We here present a coevolutionary-based method to differentially analyze subfamily specific structural features by a continuous sequence reweighting (SR) approach. We introduce the underlying principles and test its predictive capabilities on the Response Regulator family, whose subfamilies have been previously shown to display distinct, specific homo-dimerization patterns. Our results show that this reweighting scheme is effective in assigning structural features known a priori to subfamilies, even when sequence data is relatively scarce. Furthermore, sequence reweighting allows assessing if individual structural contacts pertain to specific subfamilies and it thus paves the way for the identification specificity-determining contacts from sequence variation data.

Keywords: coevolutionary analysis; direct-coupling analysis; specificity determining contacts; sequence reweighting; maximum entropy models; protein contact predictions

---

1. Introduction

The last decade has seen the emergence and maturation of coevolutionary methods aimed at predicting functionally interacting residue pairs from sequence alignments of homologous protein sequences [1–5]. The novel methodological developments, based on the use of global statistical models, have led to significant improvements in the quality of inter-residue contact prediction and computational structure prediction. This can be seen in the high scores obtained by the top-ranking teams in the recent CASP competition (Critical Assessment of Protein Structure Prediction), which mostly rely on coevolutionary predictions to guide their structural modelling [6]. Beyond the de-novo prediction of novel folds [5,7,8], coevolution-based analysis has also allowed the structural characterization of homo [2,9–11] and hetero-oligomeric [4,12,13] complexes and the determination of conformational ensembles [10,14,15].

The success of covariation-based contact prediction relies on the availability of deep multiple sequence alignments (MSAs) of homologous proteins. The rapid growth of protein sequence databases [16,17], driven by the decrease in cost of next-generation sequencing, resulted in the availability of very large protein families. In addition, the advent and improved access to meta-genomics databases further increased the pool of available sequence data [8]. In such a data-rich regime, the availability of ultra-large protein families (with typically more than 100 K homologous sequences) raises the intriguing question of how to analyze subfamily specific structural features by sequence covariation.

Indeed, coevolutionary analysis generally results in contacts predicted at the whole family level, thus predicting contact maps putatively formed by any member of the protein family. However, for large families consisting of multiple subgroups (or subfamilies), gene duplication and specialization lead to structural and functional variability of paralogous proteins sharing the overall same fold, but potentially carrying a sub-set of different contacts defining subfamily specificities [18,19]. Similarly, organisms with different genetic backgrounds and evolving in different environments will be subject to different fitness landscapes, thus not necessarily requiring the exact same structural and functional features, while still maintaining the same overall fold and function [20,21]. These observations imply that not all the members of a large protein family will necessarily satisfy all contacts predicted by coevolutionary analysis. Furthermore, the limited statistical weight of smaller subfamilies within a global alignment may prevent the identification of their specific features in a standard analysis.

This last point is of particular importance in the inspection or modelling of precise features pertaining to particular members of protein families rather than features common to the whole family. The latter scenario is typically encountered when dealing with complex eukaryotic families of great pharmacological interest, e.g., nuclear receptors [22] or G-protein coupled receptors [23], where the focus is often on specific members or sub-group. To generate predictions applicable in practice, one might need to understand or predict the effects of a small number of mutations affecting a particular set of sequences in the family. As such, novel tools and methodological developments designed for the analysis and identification of subfamily specific features at the contact level are currently required.

Some approaches have been already proposed to tackle this problem question. In [9], the authors propose to split the sequence dataset in multiple subfamily specific alignments and perform multiple independent Direct Coupling Analysis (DCA) on the sub-alignments, thereby successfully identifying subfamily specific features. Similarly, the authors in [24] performed independent DCAs on sub-class specific alignments to study different binding modes of protein complexes. Alternatively, a sequence reweighting scheme has been introduced in [10], whereby sequences belonging to two different phylogenetic groups where continuously reweighted and specific coevolutionary signals recorded, thereby assessing their phylogenetic origin. In a more recent development, authors of [25] introduced the use of restricted Boltzmann machines to simultaneously identify subfamilies and their characterizing motifs.

In the following, we build upon the reweighting concept introduced in [10], showing in a complex multi-dimensional case how this reweighting strategy can be used to identify sub-family specific contacts even in the case where the number of sequences is very low.

2. Results

To investigate how subfamily specific structural features can be extracted from complex protein families, we focus on the abundant and well-characterized family of bacterial response regulators (RR). RRs are part of the bacterial two-component signaling system, which forms one of the major transmembrane signaling systems in bacteria, and is generally composed of a transmembrane receptor and a cognate RR [26]. Upon sensing extracellular stimuli, the receptor usually auto-phosphorylates through its kinase domain and the phosphoryl group is then transferred to the RR. Prototypical RRs are generally comprised of a receiver domain, which are activated by the transfer of the phosphoryl group from the kinase domain of the receptor, and a C-terminal DNA binding domain, which acts as a transcription factor [27].

The interest in RRs as model system to study subfamily specific features originates in the fact that these proteins are classified according to their domain structure into different groups, which are characterized by alternate homo-dimerization interfaces [27]. As such, RRs form a (large) protein family, composed of several well-defined subfamilies displaying well-characterized different structural features

In particular, we focus here on three of the largest RRs subfamilies, namely the OmpR, LytTR and GerE classes, which share the same domain architecture (Figure 1A) but have been shown to exhibit different homo-dimeric arrangements. Interestingly, while RRs are classified according to the nature of the C-terminal DNA-binding domain, the receiver domain alone carries class-specific signatures, as clearly shown by a principal component analysis of its sequence space (Figure 1B). Indeed, we observe that sequences belonging to the three subfamilies form relatively well-defined clusters in the plane defined by the first two principal components, which bear the largest variance in sequence space. This finding suggests that a significant fraction of the sequence variability in the receiver domain can be explained by the nature of the tethered DNA-domain. The three subfamilies discussed here thus carry distinct sequence signatures, which pave the way for the investigation of class-specific structural features encoded in the sequence covariation of the receiver domain, as previously noted [9].

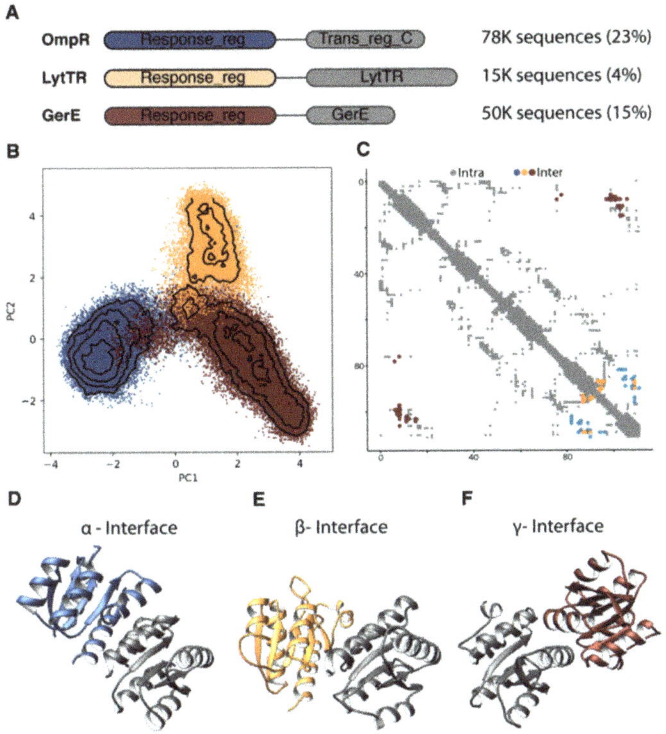

Figure 1. Sequence and Structural variability in the Response Regulator (RR) family. In all panels the color scheme follows the one defined in panel (**A**). (**A**) The three most abundant two-domain RR architectures with different dimerization modes, and their number of sequences in the complete RR alignment (fraction of the total number of sequences in parentheses). (**B**) Sequence variability of the RR family as shown by principal component projection of the RR sequences composed of the OmpR, LytTR, and GerE subfamilies. Projections are along the first two principal components. Black lines depict iso-density levels. (**C**) Contact map of three representative structures of the different subfamilies. Contacts are defined by a 5 Å distance-threshold between heavy atoms. Gray dots depict intra-molecular contacts. Colored dots depict homo-dimeric inter-molecular contacts (see Methods). (**D**–**F**) Heterogeneous homo-dimerization assemblies in the RR family. The three structural models used to define the contact map in panel C are depicted. The gray monomers in each model are structurally aligned.

The high-resolution structure of proteins belonging to these three subfamilies have been determined, and in particular, models of their homo-dimeric arrangements are available. While the overall fold of the receiver domain is very similar in all the three classes, the homo-dimeric interfaces display striking variations (Figure 1C–F). Specifically, members of the LytTR and OmpR have different arrangements but their interfaces involve similar regions of the receiver domain, as shown by the corresponding contact maps (Figure 1C). In contrast, the homo-dimeric interface of the GerE subfamily involves a dramatically different region of the contact map, clearly highlighting a completely different binding mode. Furthermore, we note that only a small set of contacts can be actually used for defining the arrangement found in the LytTR class. Indeed, several residue pairs that are involved into this homodimeric interface form structural contact at the intra-molecular level in the whole family and hence cannot be used to define a subfamily specific feature.

For the sake of clarity, we will denote hereafter the structural interfaces corresponding to the OmpR, LytTR, and GerE as α-, β-, and γ-interface, respectively (Figure 1D–F).

The most straightforward way to look for sub-class specific contacts in a DCA framework is to split the global alignment in multiple sub-class specific alignments and perform independent contact predictions. The comparative analysis of the resulting DCA predictions can then indicate which contacts are exclusively predicted in particular sets of sequences, as already shown for the RR family [9]. While effective, this approach requires that all subfamilies are composed of enough sequences to yield sufficient statistical power to perform precise contact prediction by DCA. Even if this strict requirement is fulfilled for the RR family and its subfamilies (Figure 1A), it is certainly not the case in general. We thus first investigated how the number of available sequences affects the capability of correctly assigning subfamily specific features. To this aim, we randomly subsampled the three class-specific MSAs retaining a finite fraction B_f of the original sequences and performed independent DCA predictions on these smaller alignments.

We first measured the overall prediction precision as a function of B_f, by comparing the N highest-ranked DCA predictions with a common global contact map, comprising all the intra- and inter-molecular contacts observed in the three reference structures (Figure 2A). As expected from the large size of the RR family, even subfamily alignments yield excellent overall prediction quality, with precisions of 85–90%, if we make use of all available sequences ($B_f = 1$). Remarkably, in this case the full RR alignment (union of the three subfamilies) does not yield any significant increase in precision, highlighting the probable saturation of the prediction quality. Nevertheless, decreasing the fraction of retained sequences rapidly leads to reduced precisions, and this effect is proportional to the total number of sequences in the alignment. Therefore, the gap between the results obtained for the family and those obtained for individual subfamilies initially increases for smaller B_f, while eventually the precision collapses for all the alignments when the statistical power is too low ($B_f < 0.01$). Note that at $B_f = 0.01$, the full RR alignment is comprised of 861 effective sequences (see methods), which is still acceptable for performing high-quality predictions, as suggested by the overall precision using the family alignment (76%). At the same subsampling level, the DCA results for the OmpR subfamily are still partially reliable (precision ~64%) whereas the predictions obtained with GerE and LytTR sequences are of limited to no practical use (50% and 33%, respectively).

We then specifically evaluated the range of applicability of the alignment splitting strategy for extracting subfamily features by focusing on DCA predictions of the α, β and γ interfaces. The α-interface (defined by 16 homo-dimeric contacts) is generally well recovered even at low sequence fractions using the OmpR sequences (Figure 2B). Indeed, DCA of this subfamily alignments can identify up to 60% of the contacts defining the α-interface, and roughly half of this interface is recovered on average even at $B_f = 0.01$. Reassuringly, α-interface contacts are never predicted using the GerE sequences for any subsampling even if the analysis of sufficiently large LytTR alignments yields some predictions in this interface. This result is not completely surprising, if we take into account the close proximity of the α- and the β-interfaces in the contact map. Nevertheless, the huge gap between the fractions of α-interface recovered by the two subfamily alignments makes unambiguous the assignment of the α-interface to the OmpR sequences.

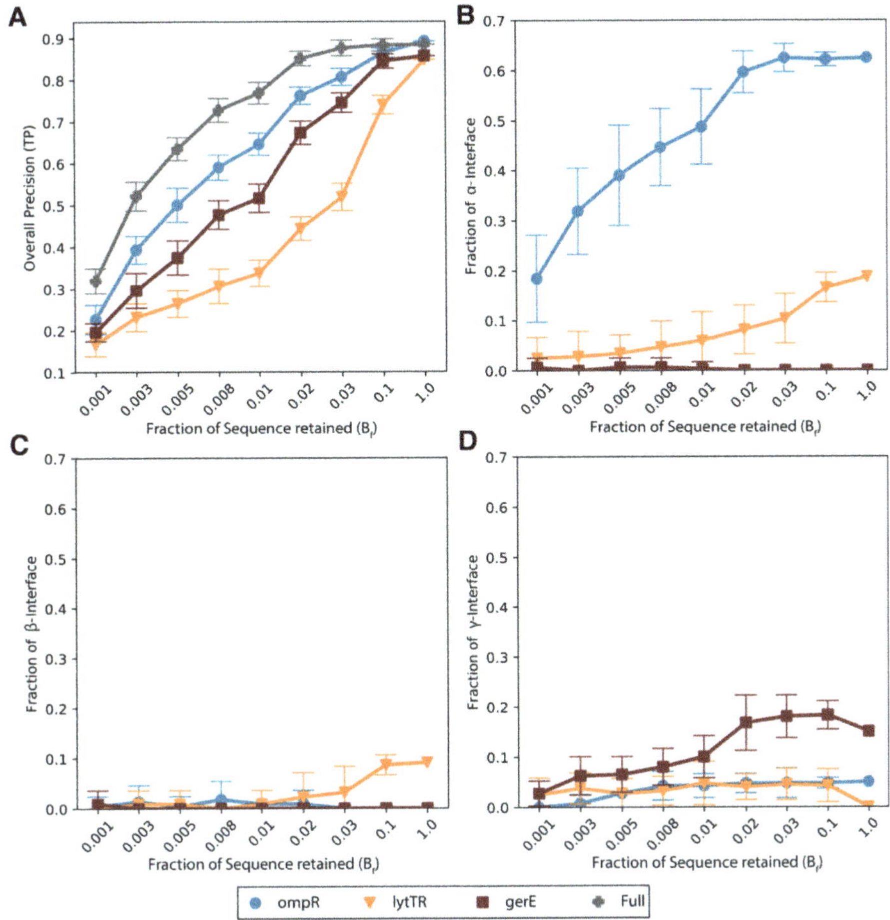

Figure 2. Prediction quality at varying alignments size. All reported quantities are shown as a function of the fraction of sequences randomly sampled from the full alignment B_f. Error bars denote standard deviations over 200 random samplings. (**A**) Overall precision (i.e., true positive rate) computed over the complete contact map (union of intra-molecular contacts and all interface contacts). Full denotes the union of all three alignments. (**B**) Fraction of the α-interface predicted in the N (112) highest ranked contacts. (**C**) Fraction of the β-interface predicted in the N (112) highest ranked contacts. (**D**) Fraction of the γ-interface predicted in the N (112) highest ranked contacts.

Conversely, the assignment of the β-interface represents a much more difficult case (Figure 2C), due to its smaller area (11 contacts) and the limited size of the cognate LytTR subfamily. In practice, even using the full alignment, only a small fraction of the interface is predicted using the LytTR sequences, whereas no β-interface contacts are predicted using either the OmpR or GerE specific alignments. The case of the γ-interface is somewhat intermediate (Figure 2D). Indeed, while this is the largest interface (20 contacts), the cognate GerE subfamily consists of significantly less sequences compared to the large OmpR. Analysis of large sub-alignments ($B_f > 0.01$) results into the prediction of ~15–20% of the γ-interface for the GerE to be compared with ~5% obtained in the case of either OmpR or LytTR sequences. This identification gap is further decreased by statistical noise as we decrease the fraction of analyzed sequences. At $B_f = 0.01$, the interface assignment to a single subfamily becomes ambiguous. At such samplings, the overall precision (Figure 2A) lies between 30% and 60%

depending on the family. Thus, the identification of subfamily specific contacts in these low sampling regimes would require dealing both with ambiguous interface assignments and with a potentially very large number of false-positive predictions even in the intra-molecular part, which lowers the overall confidence one can have in the interface predictions.

Taken together, these results illustrate that even if the splitting strategy works efficiently when sufficient sequence data is present in subfamilies [9], DCA predictions obtained with subfamily alignments might become unreliable and unable to identify subfamily specific structural features in the case of more common family sizes.

In order to circumvent this limitation, we present and discuss here an alternative scheme that does not imply the analysis of isolated subfamily alignments but instead relies on assigning arbitrary statistical weights to subfamilies within the full family alignment [10]. The core idea of this strategy is to monitor the dependence of inter-residue statistical couplings on the weights associated to subfamilies. Residue pairs whose coupling strength is strongly correlated with the weight associated to a particular subfamily will be identified as potential structural contacts specific to that subfamily. By keeping a mixture of sequences belonging to multiple sub-classes, the inference of model parameters directly controlling structural features shared by multiple subfamilies (typically intra-molecular contacts for the common fold) will benefit from increased quality of local statistics, therefore potentially helping to stabilize the overall prediction quality.

We propose the following algorithm, hereafter referred to as subfamily reweighting (SR) (see Methods for implementation details):

1. All sequences in a global alignment are subdivided into K subfamilies, indexed by k = {1, ... ,K}.
2. All the sequences belonging to a single sub-class are assigned a common weight $\omega_k \in [0, 1]$.
3. DCAs are performed, assigning weights $\{\omega_k\}$ to the sequences in the inference step, for a varying set of sequences weights.
4. The relative change in coupling scores is measured on a set of contacts of interest, as $\{\omega_k\}$ is varied.
5. Subsets of residue pairs whose overall coupling strength is strongly correlated to the change in weights are identified as subfamily specific contacts.

Additionally, we can record the overall precision computed over the whole contact map for all values of the class-specific weights. This allows identifying the regions of weight space over which the precision, taken here as proxy of our confidence in the predictions, remains in a reasonable range.

We illustrate the use of the SR approach on the RR family discussed above, in the case where only 1% of sequences are sub-sampled ($B_f = 0.01$). In this context, the SR procedure consists of the following steps: Sequences are grouped into three sub-classes corresponding to the OmpR, LytTR and GerE subfamilies. We assign all combination of weights in the range [0,1] in steps of 0.01 to the three subfamilies (see Methods) and perform DCA analysis for each set of weights. We then measure the overall precision and the coupling-scores for the α-, β-, and γ-interfaces, as a function of the subfamily weights and we report the results as triangle plots (Figure 3A–D). In this representation, the three vertices of the triangles correspond to the cases where we only keep sequences of one subfamily, while each interior point corresponds to a DCA performed with linearly interpolated weights.

We first focus on the overall precision computed over the complete contact map, which shows that the reweighting procedure results into a relatively high precision (typically above 70%) over a large range of relative weights (Figure 3A). Unsurprisingly, the quality of DCA results sharply decreases only in near vicinity of the vertexes and edges, which correspond to limiting cases where only one (vertices) or two (edges) subfamily are analyzed. In particular, the lowest precision is obtained when using only the smallest LytTR alignment (bottom-right vertex), consistently with what reported in Figure 2. Remarkably, we can explore regions of the weights-space relatively close to any extreme case while maintaining an overall precision of at least 70%, in strong contrast to the sharp precision drop obtained with subfamily alignments (Figures 2A and 3A). This finding suggests that we can reliably interpret the DCA results obtained for weights in a large portion of the weights space.

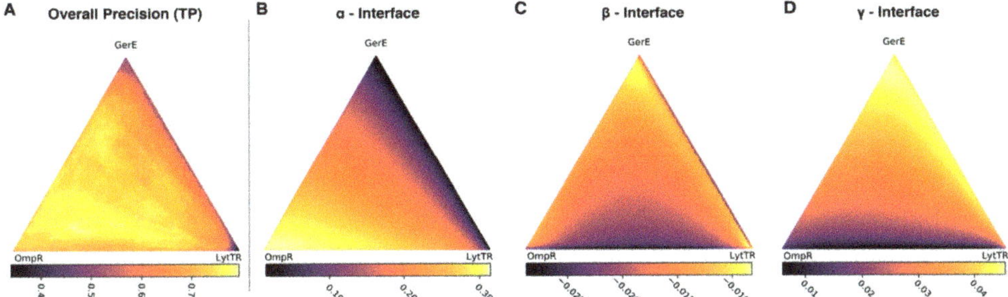

Figure 3. Results of sequence reweighting (SR). (**A**) Overall precision of the N highest ranked predictions, computed over the full contact map, comprising all intra- and inter-molecular contacts observed in the three reference structures. (**B**) Average coupling-score of the α-interface. (**C**) Average coupling-score of the β-interface. (**D**) Average coupling-score of the γ-interface.

We now inspect how the coupling-scores are affected by sequence reweighting, specifically focusing on the average coupling scores of the residue pairs defining the a priori known interfaces (Figure 3B–D) (see Methods). It appears strikingly that larger weights for the OmpR sequences correspond to higher coupling-scores over the cognate α-interface (Figure 3B). The trend is nearly linear with the orthogonal distance to the OmpR subfamily, indicating that the coevolutionary signal of the α-interface arises from the covariation encoded in OmpR sequences and it does not depend on the relative weighting of the two other subfamilies.

Interpreting the behavior of the average-coupling score over the β-interface is more difficult (Figure 3C). Indeed, there appear to be two local maxima, located both near the GerE and the cognate LytTR vertexes, with a non-monotonous behavior in the central part. This complex behavior does not lean itself to an easy interpretation and might be due to statistical noise. Indeed, the very low number of LytTR sequences in the sub-sampled alignment, combined with the relatively small number of contacts within the β-interface, may be responsible for this non-conclusive case.

In contrast, the average coupling-scores over the γ-interface display a nearly linear trend (albeit slightly tilted) with the orthogonal distance to the GerE subfamily (Figure 2D) and unambiguously identify this interface as a structural feature associated to the GerE sub-class. The clear result for the γ-interface obtained with SR approach is thus in sharp contrast with the more ambiguous assignation based on DCAs on the subfamily alignments (Figure 2D).

The success of the SR procedure in correctly characterizing both the α- and γ-interface as subfamily features, even with a limited amount of sequences, motivates us to test whether we can extend the same approach to assign individual contacts to specific subfamilies. This extension would greatly widen the range of applicability of the SR approach to protein families lacking any previous characterization of potential subfamily features.

To this aim, we can determine the coupling-scores of each residue pair as a function of the subfamily weights, analogously to what reported for whole interfaces in the triangle plots (Figure 3B–D). This information can then be used to devise scoring functions F_{ij}^k that quantify how strongly a given contact (I,j), is associated to the subfamily k (see Methods). While many functional forms can be adopted to define the scores F_{ij}^k, here we limit ourselves to a proof of principle and we test a simple approach based on multilinear kernel functions (see Methods).

We then test if this strategy may be used for associating individual homo-dimeric contacts, taken from the union of the α-, β-, and γ-interfaces, to a specific subfamily.

To this aim, we sort all the contacts using the three subfamily specific scores and we inspect for each subfamily the highest-ranked ones, which are assumed to best represent subclass features (Figure 4A–C). If we limit ourselves to the ten top-ranked pair of residues, the predictions match the cognate structural interfaces reasonably well, even using our simple functional form for the scoring functions (8/10 for α/OmpR, 5/10 for β/LytTR and 8/10 γ/GerE cases, respectively). As in previous analyses, LytTR represents the most challenging case due to the smaller amount of available sequences and the smaller cognate interface (β-interface, 11 contacts), as is reflected in the lower fraction of correctly predicted interface contacts for this subfamily (5/10 vs. 8/10).

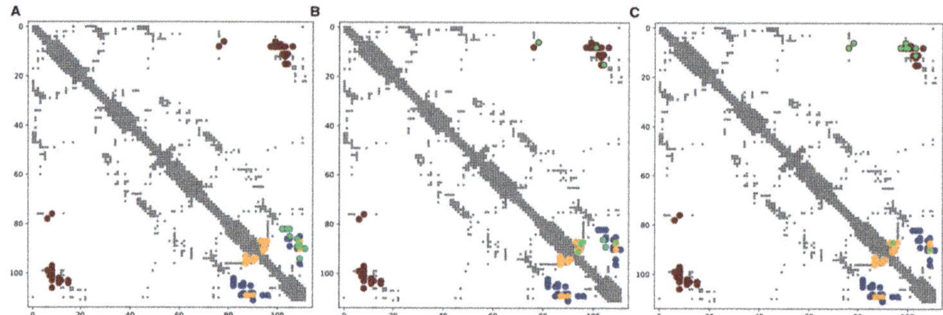

Figure 4. Identification of subfamily specific residue contacts by SR. Gray dots depict intra-molecular contacts. Colored dots depict interface contacts pertaining to the α- (blue), β- (orange) and γ- (brown-red) interfaces respectively. Dots in green are the top ranked contacts according to the F_{ij}^k scores (see Methods). (**A**) Top 10 highest ranked SR contacts for k = OmpR. (**B**) Top 10 highest ranked SR contacts for k = LytTR. (**C**) Top 10 highest ranked SR contacts for k = GerE.

These promising results, although imperfect, highlight the ability of the SR procedure to identify single residue-pairs pertaining to specific subfamilies, thus being potential candidates of specificity determining contacts. While results are based on the use of the simplest possible single-contact scoring functions as proof of principle, the use of more sophisticated subfamily specific scores will potentially increase the prediction quality of the method.

3. Discussion

The sequence reweighting approach presented here, combined with coevolutionary contact prediction, allows the characterization and analysis of pairwise contacts which pertain to protein sequences belonging to specific subfamilies. Using the well characterized response regulator family as a prototypical proof-of-concept system, we showed that the SR approach is capable of correctly assigning subfamily specific interfaces, as well as identifying specificity determining contacts. In particular, the reweighted use of all classes allows for statistically robust results, even in cases where only limited sequence data is available.

In the present work, we relied on the a-priori knowledge of the subfamilies, based on their domain architectures. In principle, this supervised component of the algorithm could be replaced by a pre-processing step consisting in clustering the sequences and thereby automatically identifying subfamilies [28–30]. In combination with the SR procedure, this would allow a large-scale search over protein sequence databases (e.g., PFAM [17]) to identify families with significant structural diversity at the subfamily level. Such an automated procedure will inevitably introduce assignment noise whose consequences on the robustness of the identification of sub-family specific contacts will have to be systematically evaluate.

Furthermore, the identification of subfamilies and their associated specific contacts might be of valuable help in the context of homology modelling, specifically in the scenario where only structural models of remote homologs are present. In such cases, it is possible to erroneously impose some structural features of the template homolog, whereas the particular target belongs to a subfamily possessing some critical structural differences. As such, being able to identify the subfamily specific contacts by sequence analysis might allow to better guide the modelling step and/or improve the critical assessment of homology models based on remote homologs.

Additionally, many structural features of protein families are currently inferred by the analysis of available "representative" structures, often determined on model organisms. While such structural models are of great value, they might actually represent "snapshots" of the heterogeneous structural ensemble characterizing the complete protein family [20]. A computational tool aiming at highlighting potential deviations from the common structural scaffold defining the whole family could thus help identifying sub-classes with novel uncharacterized structural features and thus fruitfully complement structural biology approaches.

From a practical point of view, the SR algorithm is based on associating subfamily specific weights to the sequences in the inference step. We here relied on DCA, a popular method to predict structural contacts [6], but in principle, the SR procedure can be incorporated in any prediction approach based on optimizing a data-dependent objective function analogous to the pseudo-likelihood discussed here. In a wider context, SR can be seen as an instance of a transfer-learning approach [31], whereby we make use of the available sequences of the whole protein family to maximize the statistical power of the method, while adapting it to specific sub-problems. Such classes of algorithms, aimed at maximally exploiting the available data, are of great interest particularly when the available training data is limited, as in the case of eukaryotic protein families. In this scenario, analyzing particular subfamilies requires the efficient use all the data available for the whole family, even when focusing on questions pertaining to specific paralogous sub-groups.

While we here focused on the sequence reweighting for sub-family analysis, alternative reweighting schemes have been suggested to address the problem of phylogenetic and sampling bias in large MSAs [32]. The SR procedure could in principle be easily combined with these approaches to potentially improve the prediction quality.

The identification of specificity determining features in subfamilies of protein sequences is a longstanding challenge in bioinformatics [33–35]. In this context, SR can be extended to investigate specificity determining interactions, beyond the well-studied problem of identifying specificity determining positions (SDPs). Indeed, prediction of sub-class specific contacts potentially allows the prediction of more precise structural and functional features involving pairwise epistatic interactions, non-detectable by single-site SDP analysis.

4. Materials and Methods

4.1. Sequence Data Collection and Pre-Processing

All sequence data was obtained from the PFAM database, release 32.0. We downloaded all aligned sequences for the Response regulator (RR) family (PFAM ID: PF00072), comprising 342,025 response regulator sequences of length $N = 112$ amino-acids. Additionally, the alignments for the DNA binding domains of three response regulator families where downloaded, namely the OmpR family (PFAM ID: PF00486), the LytTR family (PFAM ID: PF04397) and the GerE family (PFAM ID: PF00196).

Subfamily alignments of the response regulator domain where then built by selecting sequences from the response regulator alignment which possess either the OmpR (78,494 sequences), LytTR (14,883 sequences) or GerE (49,868 sequences) domains.

To reduce phylogenetic and sampling bias, and to simplify the reweighting procedure, the three alignments where filtered by identity, keeping only sequences in the alignments with a maximal pairwise hamming distance of 90%, using the hhfilter utility of the hhblits suite [36]. This resulted in 40,857 OmpR sequences, 12,082 LytTR and 33,344 GerE sequences. The sequences being pre-filtered by identity, the absolute number of sequences is therefore equivalent to the effective number of sequences (at a 90% identity threshold) in the current discussion.

To explore the effect of smaller datasets, sub-alignments were generated by randomly selecting a fraction B_f of sequences from the alignments. 200 random subsets where generated for analyzing the MSA depth effect (Figure 2). Three random subsets where generated for the reweighting analysis (Figures S2 and S3).

4.2. Structural Data Collection and Processing

The following structural models of the subfamily specific RR homo-dimers were collected (Table 1).

Table 1. Overview of used structural models.

Family	Interface	PDB ID	Model
OmpR	α-interface	1nxs	Biological Assembly 1
LytTR	β-interface	4cbv	Biological Assembly 1
GerE	γ-interface	4e7p	Biological Assembly 2

Inter-residue contacts were defined whenever pairs of residues had any heavy-atom distance below 5 Å. Given the close proximity of the α- and β-interfaces we chose here such a stringent contact definition threshold, lower than typically used in coevolutionary studies [2,9,37,38]. This ensures the definition of orthogonal contact interfaces. Indeed, increasing the contact threshold progressively widens the definition of interface contacts and ultimately results in partially overlapping interfaces (Figure S1).

The intra-molecular part of the contact map was defined as the union of the three intra-molecular maps from the three structural model. The three dimer interfaces were defined as all contacts not in the intra-molecular distance map for each model respectively.

In the three X-ray structures, the identification of biological assemblies is unambiguous and the corresponding homodimers do not display any noticeable break of symmetry. The three homodimers present in the asymmetric unit of 4cbv (β-interface) are characterized by the same set of intermolecular contacts according to our definition.

The contact maps were aligned to the RR multiple sequence alignment using the *mapPDB* tool from the dcaTools package [39] (available at https://gitlab.com/ducciomalinverni/dcaTools).

4.3. Direct-Coupling Analysis and Sequence Reweighting

Direct-Coupling Analysis (DCA) was performed using the asymmetric pseudo-likelihood inference method [37,40] as implemented in the lbsDCA package [39] using default inference parameters (available at https://gitlab.com/ducciomalinverni/lbsDCA). The method infers the parameters of the Hamiltonian of a generalized Potts model

$$P(s) = \frac{1}{Z} e^{\sum_{i=1}^{N} h_i(s_i) + \sum_{ij}^{N,N} J_{ij}(s_i,s_j)}$$

where $s = (s_1, \ldots, s_N)$ denotes the amino-acid sequence of length N, Z denotes the normalizing partition function and h_i and J_{ij} are model parameters controlling the single- and two-site frequencies to be fitted to the data (see, e.g., [41] for a full review on DCA and its applications).

The inference is performed by numerically minimizing the regularized negative pseudo-log likelihood l_{PL} of the data with respect to the model parameters $\{h_i, J_{ij}\}$ (see, e.g., [37] for a detailed discussion of the pseudo-likelihood inference method).

$$l_{PL} = -\frac{1}{B_{eff}} \sum_{b=1}^{B} \omega_b \, log\left(\frac{\exp\left(\sum_i h_i\left(s_i^b\right) + \sum_{i<j} J_{ij}\left(s_i^b, s_j^b\right)\right)}{\prod_i \sum_{a=1}^{21} \exp\left(h_i(a) + \sum_{j \neq i}\left(a, s_j^b\right)\right)} \right)$$

where b indexes the available sequences, ω_b denotes the weight associated to sequence b (see below), $B_{eff} = \sum_{b=1}^{B} \omega_b$ and a indexes the 21 amino-acids.

Here, we introduced subfamily specific relative weights such that $\omega_b = \omega_k$ for all sequence b belonging to subfamily $k \in \{\text{OmpR}, \text{LytTR}, \text{GerE}\}$.

We further restricted the weights to sum to unity, i.e.,

$$\omega_{\text{OmpR}} + \omega_{\text{LytTR}} + \omega_{\text{GerE}} = 1$$

While this is not strictly necessary, the normalization allows for a straightforward mapping from the 3 dimensional weights space to a visualizable 2D space. The relative weights were then varied in steps of 0.01, including the border cases $\omega_k = \{0, 1\}$.

The raw inter-residue coupling scores were computed by the Frobenius norm of the coupling parameters.

$$S_{ij} = \|J_{ij}(A, B)\|_{A,B}$$

where the norm is taken over the 20 natural amino-acids, excluding the couplings involving the gap-parameter, following [42].

Finally, the coupling scores are given by the average-product corrected (APC) raw scores following [43], i.e.,

$$\widetilde{S}_{ij} = \frac{S_{ij} - S_{i.}S_{.j}}{S_{..}}$$

where $\cdot$ denotes averaging over the relevant dimension.

All DCAs were performed using four threads per computation and took ~7 s each on a standard desktop workstation, resulting in a total computational time of roughly 10 h for generating the results presented in Figure 4.

4.4. Kernel Function Scoring

In the reweighting approach, each residue-pair is characterized by a series of coupling scores computed at different relative weights $\widetilde{S}_{ij}(\omega_{\text{OmpR}}, \omega_{\text{LytTR}}, \omega_{\text{GerE}})$. In order to annotate each contact by a single scalar for each family of interest, we introduce the following multi-linear kernel functions.

$$\varphi^k(\omega_{\text{OmpR}}, \omega_{\text{LytTR}}, \omega_{\text{GerE}}) = \begin{cases} \omega_{\text{OmpR}}(1 - \omega_{\text{LytTR}})(1 - \omega_{\text{GerE}}) & \text{if } k = \text{OmpR} \\ (1 - \omega_{\text{OmpR}})\omega_{\text{LytTR}}(1 - \omega_{\text{GerE}}) & \text{if } k = \text{LytTR} \\ (1 - \omega_{\text{OmpR}})(1 - \omega_{\text{LytTR}})\omega_{\text{GerE}} & \text{if } k = \text{GerE} \end{cases}$$

Such kernel functions have the desirable property and smoothly interpolate between these three border cases. As such, they allow to effectively compute a single weighted coupling score for each contact and each subfamily, which continuously assigns higher weights to coupling scores computed in realizations which weighted sequences of the subfamily more.

$$\varphi^k = \begin{cases} 1 & \text{if } \omega_k = 1 \\ 0 & \text{if } \omega_i = 1 \; \forall \; i \neq k \end{cases}$$

In practice, to focus on the relative variation of coupling scores, irrespective of their absolute value, we subtract for each residue pair the average coupling score $< \widetilde{S}_{ij} >$ (averaged over all weights triplets) before computing the kernel integral. This finally allows to define an average-corrected subfamily specific score for each residue-pair.

$$F_{ij}^k = \sum_{\omega_{\text{OmpR}}, \omega_{\text{LytTR}}, \omega_{\text{GerE}}} \varphi^k(\omega_{\text{OmpR}}, \omega_{\text{LytTR}}, \omega_{\text{GerE}})(\widetilde{S}_{ij}(\omega_{\text{OmpR}}, \omega_{\text{LytTR}}, \omega_{\text{GerE}}) - < \widetilde{S}_{ij} >)$$
$$\forall \; k \in \{\text{OmpR, LytTR, GerE}\}$$

Supplementary Materials: The following are available online at http://www.mdpi.com/1099-4300/21/11/1127/s1, Figure S1: Structural contact maps and overlap between structural interfaces at increasing contact threshold.

Gray dots depict intra-molecular contacts. Colored dots depict homo-dimeric inter-molecular contacts. Figure S2: Triangle plots for the SR procedure for the three random sub-samplings (analog to Figure 3). Figure S3: Identification of subfamily specific residue contacts by SR (analog to Figure 4), for all three random sub-samplings. A Top 10 highest ranked SR contacts for k = OmpR, replica 1. B Top 10 highest ranked SR contacts for k = LytTR, replica 1. C Top 10 highest ranked SR contacts for k = GerE, replica 1. D Top 10 highest ranked SR contacts for k = OmpR, replica 2. E Top 10 highest ranked SR contacts for k = LytTR, replica 2. F Top 10 highest ranked SR contacts for k = GerE, replica 2. G Top 10 highest ranked SR contacts for k = OmpR, replica 3. H Top 10 highest ranked SR contacts for k = LytTR, replica 3. I Top 10 highest ranked SR contacts for k = GerE, replica 3.

Author Contributions: Conceptualization, D.M. and A.B.; Methodology, D.M. and A.B.; Formal analysis, D.M. and A.B.; Investigation, D.M. and A.B.; Resources, D.M. and A.B.; Writing—original draft preparation, D.M. and A.B.; Writing—review and editing, D.M. and A.B.; Visualization, D.M. and A.B.

Funding: This research was funded by the Swiss National Science Foundation (SNFS), grant number P2ELP3_181910 (DM) and French Agence Nationale de la Recherche (ANR), under grant ANR-14-ACHN-0016 (AB).

Acknowledgments: We thank Paolo De Los Rios for useful discussions regarding the SR procedure and Stefano Trapani for helping us with the analysis of X-ray structures.

Conflicts of Interest: The authors declare no conflict of interest.

References

1. Weigt, M.; White, R.A.; Szurmant, H.; Hoch, J.A.; Hwa, T. Identification of direct residue contacts in protein-protein interaction by message passing. *Proc. Natl. Acad. Sci. USA* **2009**, *106*, 67–72. [CrossRef] [PubMed]

2. Morcos, F.; Pagnani, A.; Lunt, B.; Bertolino, A.; Marks, D.S.; Sander, C.; Zecchina, R.; Onuchic, J.N.; Hwa, T.; Weigt, M. Direct-coupling analysis of residue coevolution captures native contacts across many protein families. *Proc. Natl. Acad. Sci. USA* **2011**, *108*, E1293–E1301. [CrossRef] [PubMed]

3. Jones, D.T.; Buchan, D.W.A.; Cozzetto, D.; Pontil, M. PSICOV: Precise structural contact prediction using sparse inverse covariance estimation on large multiple sequence alignments. *Bioinformatics* **2012**, *28*, 184–190. [CrossRef] [PubMed]

4. Ovchinnikov, S.; Kamisetty, H.; Baker, D. Robust and accurate prediction of residue-residue interactions across protein interfaces using evolutionary information. *Elife* **2014**, *3*, e02030. [CrossRef] [PubMed]

5. Marks, D.S.; Colwell, L.J.; Sheridan, R.; Hopf, T.A.; Pagnani, A.; Zecchina, R.; Sander, C. Protein 3D structure computed from evolutionary sequence variation. *PLoS ONE* **2011**, *6*, e28766. [CrossRef] [PubMed]

6. Schaarschmidt, J.; Monastyrskyy, B.; Kryshtafovych, A.; Bonvin, A.M.J.J. Assessment of contact predictions in CASP12: Co-evolution and deep learning coming of age. *Proteins Struct. Funct. Bioinform.* **2018**, *86*, 51–66. [CrossRef] [PubMed]

7. Hopf, T.A.; Colwell, L.J.; Sheridan, R.; Rost, B.; Sander, C.; Marks, D.S. Three-dimensional structures of membrane proteins from genomic sequencing. *Cell* **2012**, *149*, 1607–1621. [CrossRef] [PubMed]

8. Ovchinnikov, S.; Park, H.; Varghese, N.; Huang, P.-S.; Pavlopoulos, G.A.; Kim, D.E.; Kamisetty, H.; Kyrpides, N.C.; Baker, D. Protein structure determination using metagenome sequence data. *Science* **2017**, *355*, 294–298. [CrossRef]

9. Uguzzoni, G.; John Lovis, S.; Oteri, F.; Schug, A.; Szurmant, H.; Weigt, M. Large-scale identification of coevolution signals across homo-oligomeric protein interfaces by direct coupling analysis. *Proc. Natl. Acad. Sci. USA* **2017**, *114*, E2662–E2671. [CrossRef]

10. Malinverni, D.; Marsili, S.; Barducci, A.; De Los Rios, P. Large-Scale Conformational Transitions and Dimerization Are Encoded in the Amino-Acid Sequences of Hsp70 Chaperones. *PLoS Comput. Biol.* **2015**, *11*, e1004262. [CrossRef]

11. Fantini, M.; Malinverni, D.; De Los Rios, P.; Pastore, A. New Techniques for Ancient Proteins: Direct Coupling Analysis Applied on Proteins Involved in Iron Sulfur Cluster Biogenesis. *Front. Mol. Biosci.* **2017**, *4*, 1–14. [CrossRef] [PubMed]

12. Hopf, T.A.; Schärfe, C.P.I.; Rodrigues, J.P.G.L.M.; Green, A.G.; Kohlbacher, O.; Sander, C.; Bonvin, A.M.J.J.; Marks, D.S. Sequence co-evolution gives 3D contacts and structures of protein complexes. *Elife* **2014**, *3*, e03430. [CrossRef] [PubMed]

13. Malinverni, D.; Lopez, A.J.; Rios, P.D.L.; Hummer, G.; Barducci, A. Modeling Hsp70/Hsp40 interaction by multi-scale molecular simulations and co-evolutionary sequence analysis. *Elife* **2016**, 1–17. [CrossRef]

14. Sutto, L.; Marsili, S.; Valencia, A.; Gervasio, F.L. From residue coevolution to protein conformational ensembles and functional dynamics. *Proc. Natl. Acad. Sci. USA* **2015**, *112*, 13567–13572. [CrossRef] [PubMed]

15. Morcos, F.; Jana, B.; Hwa, T.; Onuchic, J.N. Coevolutionary signals across protein lineages help capture multiple protein conformations. *Proc. Natl. Acad. Sci. USA* **2013**, *110*, 20533–20538. [CrossRef] [PubMed]

16. Bateman, A.; Martin, M.J.; O'Donovan, C.; Magrane, M.; Alpi, E.; Antunes, R.; Bely, B.; Bingley, M.; Bonilla, C.; Britto, R.; et al. UniProt: The universal protein knowledgebase. *Nucleic Acids Res.* **2017**, *45*, D158–D169.

17. Finn, R.D.; Mistry, J.; Tate, J.; Coggill, P.; Heger, A.; Pollington, J.E.; Gavin, O.L.; Gunasekaran, P.; Ceric, G.; Forslund, K.; et al. The Pfam protein families database. *Nucleic Acids Res.* **2010**, *38*, D211–D222. [CrossRef]

18. Marchant, A.; Cisneros, A.F.; Dubé, A.K.; Gagnon-Arsenault, I.; Ascencio, D.; Jain, H.; Aubé, S.; Eberlein, C.; Evans-Yamamoto, D.; Yachie, N.; et al. The role of structural pleiotropy and regulatory evolution in the retention of heteromers of paralogs. *Elife* **2019**, *8*, 1–34. [CrossRef]

19. Peterson, M.E.; Chen, F.; Saven, J.G.; Roos, D.S.; Babbitt, P.C.; Sali, A. Evolutionary constraints on structural similarity in orthologs and paralogs. *Protein Sci.* **2009**, *18*, 1306–1315. [CrossRef]

20. Chothia, C.; Lesk, A.M. The relation between the divergence of sequence and structure in proteins. *Embo J.* **1986**, *5*, 823–826. [CrossRef]

21. Anishchenko, I.; Ovchinnikov, S.; Kamisetty, H.; Baker, D. Origins of coevolution between residues distant in protein 3D structures. *Proc. Natl. Acad. Sci. USA* **2017**, *114*, 9122–9127. [CrossRef] [PubMed]

22. Robinson-Rechavi, M.; Escriva, H.; Laudet, V. The nuclear receptor superfamily. *J. Cell Sci.* **2003**, *116*, 585–586. [CrossRef] [PubMed]

23. Hauser, A.S.; Attwood, M.M.; Rask-Andersen, M.; Schiöth, H.B.; Gloriam, D.E. Trends in GPCR drug discovery: New agents, targets and indications. *Nat. Rev. Drug Discov.* **2017**, *16*, 829–842. [CrossRef] [PubMed]

24. Nillegoda, N.B.; Stank, A.; Malinverni, D.; Alberts, N.; Szlachcic, A.; Barducci, A.; De Los Rios, P.; Wade, R.C.; Bukau, B. Evolution of an intricate J-protein network driving protein disaggregation in eukaryotes. *Elife* **2017**, *6*, e24560. [CrossRef] [PubMed]

25. Tubiana, J.; Cocco, S.; Monasson, R. Learning protein constitutive motifs from sequence data. *Elife* **2019**, *8*, e39397. [CrossRef]

26. Jung, K.; Fabiani, F.; Hoyer, E.; Lassak, J. Bacterial transmembrane signalling systems and their engineering for biosensing. *Open Biol.* **2018**, *8*. [CrossRef]

27. Zschiedrich, C.P.; Keidel, V.; Szurmant, H. Molecular mechanisms of two-component signal transduction. *J. Mol. Biol.* **2016**, *428*, 372–3775. [CrossRef]

28. Steinegger, M.; Söding, J. Clustering huge protein sequence sets in linear time. *Nat. Commun.* **2018**, *9*, 2542. [CrossRef]

29. Chen, Y.; Reilly, K.D.; Sprague, A.P.; Guan, Z. Seqoptics: A protein sequence clustering method. In Proceedings of the First International Multi-Symposiums on Computer and Computational Sciences (IMSCCS'06), Hangzhou, China, 20–24 June 2006; pp. 69–75.

30. Fu, L.; Niu, B.; Zhu, Z.; Wu, S.; Li, W. CD-HIT: Accelerated for clustering the next-generation sequencing data. *Bioinformatics* **2012**, *28*, 3150–3152. [CrossRef]

31. Yang, Q.; Pan, S.J. A Survey on Transfer Learning. *IEEE Trans. Knowl. Data Eng.* **2010**, *22*, 1345–1359.

32. Hockenberry, A.J.; Wilke, C.O. Phylogenetic weighting does little to improve the accuracy of evolutionary coupling analyses. *Entropy* **2019**, *21*, 1000. [CrossRef] [PubMed]

33. Mirny, L.A.; Gelfand, M.S. Using orthologous and paralogous proteins to identify specificity-determining residues in bacterial transcription factors. *J. Mol. Biol.* **2002**, *321*, 7–20. [CrossRef]

34. Chakraborty, A.; Chakrabarti, S. A survey on prediction of specificity-determining sites in proteins. *Brief. Bioinform.* **2013**, *16*, 71–88. [CrossRef] [PubMed]

35. Sloutsky, R.; Naegle, K.M. High-resolution identification of specificity determining positions in the LacI protein family using ensembles of sub-sampled alignments. *PLoS ONE* **2016**, *11*, 1–21. [CrossRef]

36. Remmert, M.; Biegert, A.; Hauser, A.; Söding, J. HHblits: Lightning-fast iterative protein sequence searching by HMM-HMM alignment. *Nat. Methods* **2012**, *9*, 173–175. [CrossRef]

37. Ekeberg, M.; Lövkvist, C.; Lan, Y.; Weigt, M.; Aurell, E. Improved contact prediction in proteins: Using pseudolikelihoods to infer Potts models. *Phys. Rev. E* **2013**, *87*, 1–16. [CrossRef]

38. Hockenberry, A.J.; Wilke, C.O. Evolutionary couplings detect side-chain interactions. *PeerJ* **2019**, *7*, e7280. [CrossRef]

39. Malinverni, D.; Barducci, A. Coevolutionary Analysis of Protein Sequences for Molecular Modeling. In *Biomolecular Simulations: Methods and Protocols*; Bonomi, M., Camilloni, C., Eds.; Springer: New York, NY, USA, 2019; pp. 379–397. ISBN 978-1-4939-9608-7.

40. Ekeberg, M.; Hartonen, T.; Aurell, E. Fast pseudolikelihood maximization for direct-coupling analysis of protein structure from many homologous amino-acid sequences. *J. Comput. Phys.* **2014**, *276*, 341–356. [CrossRef]

41. Cocco, S.; Feinauer, C.; Figliuzzi, M.; Monasson, R.; Weigt, M. Inverse statistical physics of protein sequences: A key issues review. *Rep. Prog. Phys.* **2018**, *81*, 9965. [CrossRef]

42. Feinauer, C.; Skwark, M.J.; Pagnani, A.; Aurell, E. Improving Contact Prediction along Three Dimensions. *PLoS Comput. Biol.* **2014**, *10*, e1003847. [CrossRef]

43. Dunn, S.D.; Wahl, L.M.; Gloor, G.B. Mutual information without the influence of phylogeny or entropy dramatically improves residue contact prediction. *Bioinformatics* **2008**, *24*, 333–340. [CrossRef] [PubMed]

Article

Allostery and Epistasis: Emergent Properties of Anisotropic Networks

Paul Campitelli and S. Banu Ozkan *

Department of Physics and Center for Biological Physics, Arizona State University, Tempe, AZ 85287, USA;
Paul.Campitelli@asu.edu
* Correspondence: Banu.Ozkan@asu.edu

Received: 24 April 2020; Accepted: 8 June 2020; Published: 16 June 2020

Abstract: Understanding the underlying mechanisms behind protein allostery and non-additivity of substitution outcomes (i.e., epistasis) is critical when attempting to predict the functional impact of mutations, particularly at non-conserved sites. In an effort to model these two biological properties, we extend the framework of our metric to calculate dynamic coupling between residues, the Dynamic Coupling Index (DCI) to two new metrics: (i) EpiScore, which quantifies the difference between the residue fluctuation response of a functional site when two other positions are perturbed with random Brownian kicks simultaneously versus individually to capture the degree of cooperativity of these two other positions in modulating the dynamics of the functional site and (ii) DCI_{asym}, which measures the degree of asymmetry between the residue fluctuation response of two sites when one or the other is perturbed with a random force. Applied to four independent systems, we successfully show that EpiScore and DCI_{asym} can capture important biophysical properties in dual mutant substitution outcomes. We propose that allosteric regulation and the mechanisms underlying non-additive amino acid substitution outcomes (i.e., epistasis) can be understood as emergent properties of an anisotropic network of interactions where the inclusion of the full network of interactions is critical for accurate modeling. Consequently, mutations which drive towards a new function may require a fine balance between functional site asymmetry and strength of dynamic coupling with the functional sites. These two tools will provide mechanistic insight into both understanding and predicting the outcome of dual mutations.

Keywords: epistasis; allostery; elastic network model; protein conformational dynamics

---

1. Introduction

A growing body of data on the human genome suggest that within the exome (the protein coding region), one individual may possess 10,000 or more non-synonymous nucleotide variants, many of which occur at positions which are not evolutionarily conserved [1–3]. Predicting the functional outcome of mutations at non-conserved sites remains an extremely difficult challenge. In particular, providing accurate predictions about the impact of these variations is difficult when only considering single, independent point mutations without accounting for the background of other positions and their chemical specificity (i.e., context dependence).

One reason why predicting the impact of mutations may fail is that extensive epistasis occurs during evolution [4–6]. Epistasis is defined as a context-dependent functional outcome, where, the alternative context could be just one single amino acid difference, or it could be a paralog with 25% sequence identity. Experimentally, epistasis manifests as a non-additive outcome from two or more amino acid changes within a protein. The effects can be dramatic. For example, a substitution may only confer a beneficial effect upon fixation of a second-site, also known as a "permissive" change; conversely, a neutral substitution might become deleterious in the presence of other "restrictive" substitutions [7–9]. Thus, epistasis plays a vital role in shaping trajectories of protein evolution [4–10]. Furthermore, mounting evidence indicates that protein evolution, particularly evolution towards new function, proceeds not only through mutations at functionally critical sites, but also through sites which can have a subtle (or, occasionally, substantial) effect on function when mutated without being immediately identifiable as positions with particular functional or structural importance [11–13].

Epistatic relationships becomes crucial when comparing homologous protein families or protein domains, which can exhibit significant sequence variation and biochemical properties that may span orders of magnitude while still maintaining a similar three-dimensional (3-D) fold [14–19]. Thus, single or dual mutations on homologous proteins yield a wide range of functional outcomes [20–24]. In fact, understanding the mechanics or predicting the results of dual mutations remains a significant challenge in the presence of systems which experience large epistatic effects, even when accurate experimental data are available for the single mutant systems [7–10,25].

On the other hand, when protein equilibrium dynamics and each individual position's contribution to these dynamics are taken into consideration, we can shed light onto the mechanism of epistatic relations. This is because proteins sample many different conformations within the native state, and these conformational dynamics, governed by the strength of the 3-D network of interactions, underlie protein function. Within this dynamic view, we can simply treat a protein as a biological signal processor where the 3-D interaction network mediates long-range communication through amino acid fluctuations nascent to a given protein sequence. Therefore, the knowledge of how mutations may fine-tune this sequence-function relationship necessitates evaluating the role of each residue position in establishing a protein's internal communication network through protein dynamics [26,27]. Particularly, when two substitutional sites are considered together, the dynamic coupling of these sites results in a joint effect (i.e., a cooperative response) leading to the modulation of signal processing responsible for biophysical behavior and, ultimately, may give rise to a non-additive functional outcome.

The non-additive, epistatic interactions therefore can use dynamic features of a protein to modulate function. These dynamics features are similar to that found in allosteric modulation in which a protein is able to control catalytic function or regulate on/off states through the binding of a ligand to a site distal from a catalytic/active site. This distal binding has been shown to modulate catalytic site dynamics, sometimes without association to distinct conformational states. This type of allostery, which can impact function by manipulating the normal modes of the protein while retaining the conformation, is known as dynamic allostery [28,29]. We now understand this form of allosteric regulation to be a specific, and often more dramatic, emergent property of the unique internal networking between amino acids within a protein. To this end, allosteric systems reduce the enormous dimensionality associated with information transfer and communication pathways for these complex, anisotropic networks by identifying important regulation sites a priori. Therefore, as observed in allosteric regulations, the long-distance interactions through dynamic coupling between different positions and active sites can be modulated and re-wired through substitutions, which emerge as epistasis that drives the evolution of new function. Here, we aim to identify these epistatic relations through the development of dynamics-based metrics which can measure the strength of long-range dynamic interactions.

The modeling of protein conformational dynamics using force perturbations and elastic networks has been previously used successfully in attempts to understand the role of long-range interactions in protein evolution [30–33]. Here we attempt to model these effects through the use of Perturbation Response Scanning (PRS) and the Elastic Network Model (ENM) to construct a Dynamic Coupling Index (DCI) where we can capture the dynamic coupling between any given residue pair or set of residues via a system's response to random force perturbations. DCI captures the strength of displacement response of a given position i upon perturbation to a single position (or subset of positions) j, relative to the average fluctuation response of position i when all of the positions within a structure are perturbed. Expanding upon the dynamic coupling concept, here we develop a new metric called EpiScore. EpiScore measures the difference in the residue fluctuation response of an active site when two mutational sites are simultaneously perturbed by random forces versus the response when individual force perturbations are exerted one at a time to the mutational sites.

In order to determine whether EpiScore can identify the degree and strength of epistatic relationships between position pairs, we first applied our analysis to the deep scanning database of double mutations between all positions in the IgG-binding domain of protein GB1 [24]. These modern, high-throughput screens (e.g., deep mutational scans) assay large numbers of mutants (up to 10^8) [34–37], but the information is largely qualitative. Therefore, to further test whether our approach can identify epistatic relations which specifically emerge during the evolution of new function, we applied our methodology to two different protein systems where the traditional biochemical quantifications of mutational effects (e.g., k_{cat}, K_M, IC_{50}) for a range of substrates are available. These two systems, *P. falciparum* DHFR (pfDHFR) and a β-lactamase (TEM-1), naturally confer resistance to drugs and the trajectories of these resistances as well as their epistatic relationships have been explored [22,23]. Importantly, these two systems are also known to be allosteric proteins.

We first observed that EpiScore can distinguish positive and negative epistasis in dual mutations when analysis was performed over 1045 single mutants and 509,963 double mutants of GB1. We also found that the average EpiScore value correlates well with experimental epistatic measures calculated using pyrimethamine IC_{50} values of pfDHFR dual mutants and the catalytic turnover rates for cefotaxime of TEM-1 dual mutants. Furthermore, each pfDHFR amino acid pair exhibits distinct distributions of EpiScore values showing the importance of how these two positions communicate with the active site through the anisotropic interaction network.

Interestingly, DCI is usually not symmetric, i.e., the fluctuation response of position *i* upon exerting random forces on *j* is not identical to the response of *j* when position *i* is perturbed; we calculate this asymmetry with DCI_{asym}. We applied our DCI_{asym} analysis to the TEM-1 dual-mutant sites and found that, indeed, a relationship exists between dynamic coupling asymmetry and EpiScore when all active sites in the TEM-1 system are considered. Specifically, two of the three dual mutant positions which exhibited the largest positive epistasis in cefotaxime k_{cat}/K_M from the wild-type had both EpiScore values < 1 (indicating strong non-additivity) with respect to active site S70. Additionally, these dual mutants also exhibit asymmetry in dynamic coupling based on DCI_{asym}, with consistent unidirectionality from active sites site to mutation sites in long range communication. We propose that this communication directionality signature should be readily apparent in known allosteric systems as mentioned above. Therefore, we applied a similar analysis to a Pin1 protein well-studied for its dynamic allostery and showed that the DCI_{asym} between the catalytic binding sites and non-catalytic distal binding sites presents a unique directionality in long distance dynamic coupling, leading to a cause-and-effect relationship between allosteric sites and active sites also observed in epistatic interactions.

2. Methods

We previously designed a unique way to capture site-specific coupling between residue pairs or groups of residues, the Dynamic Coupling Index (DCI). The underlying premise behind DCI is the importance of a system's response to a force perturbation, be that protein-solvent, protein-protein, protein-ion or protein ligand interactions. Additionally, the point mutations here are modeled by the response of a system to a perturbation at a specific site, a.k.a. a single amino acid.

DCI is a combination of the Elastic Network Model (ENM) and Linear Response Theory (LRT) where the protein is modeled by representing the amino acids as nodes in a network connected by Hookean springs (Figure 1). The interaction between two amino acids close in space due to their 3-dimensional structure is represented by a simple harmonic function. A random Brownian kick in the form of a unit force perturbation is applied to an individual position which generates a response propagating through the rest of the structure, causing other positions to respond to this perturbation through the network of interactions. Using LRT, we can calculate the fluctuation response ΔR (Equation (1)) of each position and create response vector that measures the magnitude and direction (x, y and z) of displacement of every residue from its mean. As mentioned above, this (to the first order) mimics the effects of in vivo interactions of a protein. For example, a ligand binding event will apply a force to residues in the binding pocket of a receptor protein. In our perturbation residue scanning (PRS) approach, this is averaged over many unit force directions to simulate an isotropic perturbation.

$$[\Delta \mathbf{R}]_{3N\times1} = [\mathbf{H}]_{3N\times3N}^{-1}\,[\mathbf{F}]_{3N\times1} \tag{1}$$

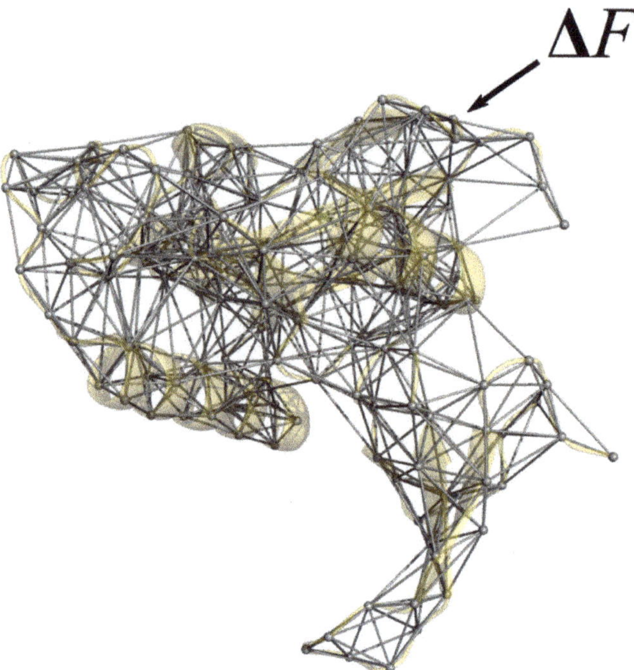

Figure 1. Elastic network model representation of protein Pin1 (PDB ID 3TCZ [7], ligands removed). Here, each residue within the structure is represented as a single node at the Cα position, connected to other nodes via Hookean springs. Using a combination of Perturbation Response Scanning (PRS) and Linear Response Theory (LRT) [38,39], each residue is perturbed by a Brownian kick applied as an isotropic external force which then generates a fluctuation response in all other residues within the network. This figure was rendered in PyMol [40].

H is the Hessian, a 3N × 3N matrix which can be constructed from 3-dimensional atomic coordinate information where it is composed of the second order derivatives of the harmonic potential energy with respect to the components of the position vector of length 3N. The Hessian matrix can be extracted directly from molecular dynamics simulations as the inverse of the covariance matrix. This method allows one to implicitly capture specific physiochemical properties and more accurate residue-residue interactions via atomistic force fields and subsequent all-atom simulation data. However, for the purposes of this paper, we wished to investigate only those relationships which could be derived solely from inter-atomic distances of single protein structures and thus we used the ENM version of our approach.

Repeating this process, each position in the structure is perturbed sequentially to generate a perturbation response matrix **A**

$$\mathbf{A}_{N \times N} = \begin{bmatrix} \left|\Delta R^1\right|_1 & \cdots & \left|\Delta R^N\right|_1 \\ \vdots & \ddots & \vdots \\ \left|\Delta R^1\right|_N & \cdots & \left|\Delta R^N\right|_N \end{bmatrix} \tag{2}$$

where $\left|\Delta R^j\right|_i = \sqrt{\langle (\Delta R)^2 \rangle}$ is the magnitude of fluctuation response at position i due to the perturbations at position j. From this perturbation response matrix, we can construct DCI. DCI_{ij}, then, represents the displacement response of position i upon perturbation to a single functionally important position (or subset of positions) j, relative to the average fluctuation response of position i when all of the positions within a structure are perturbed.

$$DCI_{ij} = \frac{\sum_j^{N_{functional}} \left|\Delta R^j\right|_i / N_{functional}}{\sum_{j=1}^{N} \left|\Delta R^j\right|_i / N} \tag{3}$$

As such, DCI can be considered a measure of the dynamic coupling between residue i and residue(s) j upon perturbation to residue(s) j.

It is often more convenient to represent DCI as a percentile rank,

$$\%DCI_{ij} = \frac{m_{\leq i}}{N} \tag{4}$$

where $m_{\leq i}$ is the number of positions with a DCI value $\leq DCI_{ij}$ for a system of N residues.

One of the most important aspects of DCI is that the entire network of interactions is explicitly included in subsequent calculations without the need of dimensionality reduction such as Normal Mode Analysis through principal component analysis. If one considers interactions such as allostery as an emergent property of an anisotropic interaction network, it is critical to include the interactions of the entire network to accurately model the effect one residue can have on another.

Here, we present two further extensions of DCI which allow us to uniquely model allosteric interactions and epistatic effects; EpiScore and DCI_{asym}, respectively. EpiScore can identify or describe potential non-additivity in substitution behavior between residue pairs. This metric can capture the differences in a normalized perturbation response to a position k when a force is applied at two residues i and j simultaneously versus the average additive perturbation response when each residue i, j, is perturbed individually (Figure 2). EpiScore values < 1 (>1) indicate that the additive perturbations of positions i and j generates a greater (lesser) response at position k than the effect of a simultaneous perturbation. This means that, when treated together with a simultaneous perturbation at both sites i and j, the displacement response of k is lower (higher) as compared to the average effect of individual perturbations to i and j, one at a time. As EpiScore is a linear scale, the further the value from 1, the greater the effect described above.

Interestingly, through the use of DCI we can capture asymmetry between different residues within a protein, as coupling in and of itself is asymmetric within an anisotropic network. That is, each amino acid has a set of positions to which it is highly coupled, and this anisotropy in connections gives rise to unique differences in coupling between a given i j pair of amino acids which do not have direct interactions (Figure 3). DCI_{asym}, then, is simply DCI_{ij} (the normalized displacement response of position j upon a perturbation to position i) $- DCI_{ji}$ (Equation (5)). Using DCI_{asym} we can determine a cause-effect relationship between the i j pair in terms of force/signal propagation between these two positions.

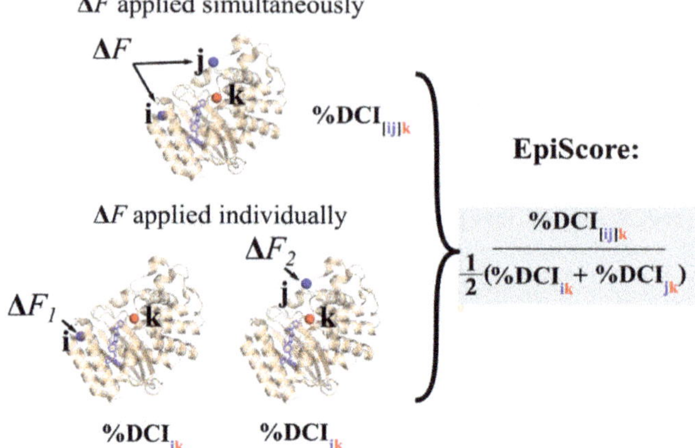

Figure 2. Schematic describing the calculation of EpiScore. The numerator is the %Dynamic Coupling Index (%DCI) value at position k upon a simultaneous perturbation to positions i and j divided by the average %DCI value at position k when positions i and j are perturbed individually. Thus, an EpiScore value of 1 indicates a perfect coupling additivity with respect to a given position k in individual versus simultaneous perturbations of two positions i and j. Figures rendered in PyMol [40] using β-lactamase (TEM-1) structure 1BTL [41].

$$DCI_{asym} = DCI_i - DCI_j \tag{5}$$

$$\%DCI_{asym} = \%DCI_i - \%DCI_j \tag{6}$$

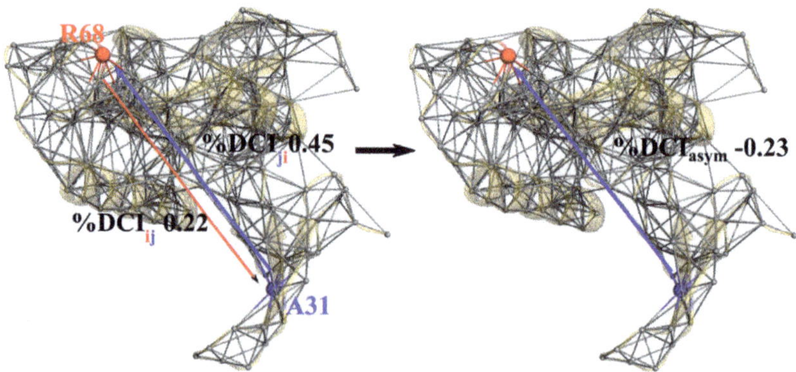

Figure 3. Example of asymmetric coupling between residue R68 of the PPIase domain and A31 of the WW domain in Pin1 (PDB ID 3TCZ [7]). The differences in local contacts give rise to network inhomogeneities which subsequently result in different %DCI values from R68 to A31 versus A31 to R68 (left). The subtraction of these two values gives a measure of coupling directionality upon external perturbations between these two sites (right). These figures were rendered in PyMol [40].

3. Results and Discussion

3.1. Epistasis and EpiScore

To investigate the relationship between internal networking and epistasis, we first apply our analysis to protein G domain B1 (GB1, PDB ID 2QMT [42]), for which there exists a comprehensive set of mutational data. Specifically, fitness effects of mutations were determined with high confidence for

1045 single mutants and 509,963 double mutants, with data available for all 1485 possible position pairs [24]. In this work, experimental epistasis was calculated as $\ln(W_{ab}) - \ln(W_a) - \ln(W_b)$, where W_{ab} represents the fitness for the dual mutant and W_a and W_b are the fitness values for the single mutants. Here we investigate the relationship between the experimental epistasis and EpiScore by comparing the average EpiScore for each position pair with instances of positive (blue) or negative (red) epistasis using the skewness of the experimental epistasis distribution over the full mutational space available for a given pair (Figure 4). Skewness was chosen as it more accurately represented the substitution behavior than position averages, which would often tend towards zero without capturing the substitution behavior for a given position pair. EpiScore values were calculated for all position pairs relative to every other position within the protein and averaged over, generating one average EpiScore value for each pair. Interestingly, when we obtained the average EpiScore distribution of experimental positive and negative epistatic pairs we found that EpiScore values above and below one tend to distinctly divide positive from negative epistasis; positive experimental epistasis was more frequently skewed towards EpiScore > 1, and likewise negative cases are skewed towards EpiScore < 1.

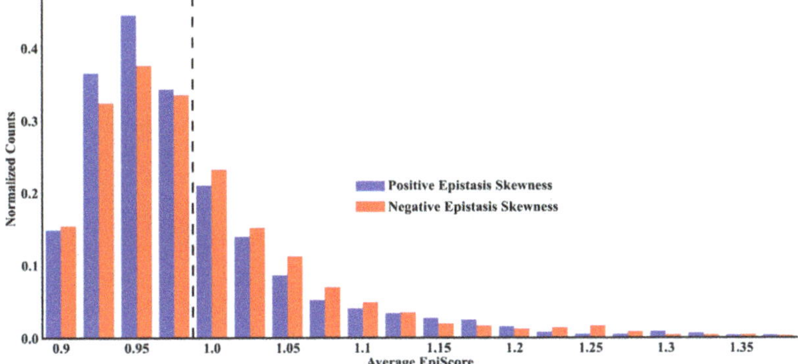

Figure 4. Distribution of the average EpiScore for protein GB1 protein pairs separated by positive and negative experimental epistasis using experimental deep scanning data for every position pair (excluding position 1). EpiScore values above and below one (dashed line) tend to distinctly divide cases for which experimental epistasis was more frequently skewed towards the positive (below one) and negative (above one)).

The full system analysis of GB1 showed the existence of a general trend between epistasis and EpiScore; particularly, an inverse relationship between EpiScore above or below one and skewness in experimental epistasis, indicating that positions with EpiScore less than 1 more often work cooperatively towards beneficial protein function, whereas pairs yielding EpiScore values greater than 1 usually result in antagonistic interactions which impair function. In an effort to elucidate more specific mechanistic details or trends underlying epistatic interactions which may exist in other systems, we broaden our application of EpiScore to other known epistatic proteins with a focus on specific mutation pairs. As such, we next study DHFR, a protein involved in the development of anti-malarial resistances in malarial parasites. Anti-malarial drugs commonly target the DHFR, which catalyzes the reduction of dihydrofolate and is essential to cellular growth and proliferation. Pyrimethamine is one such drug, used to treat malaria caused by one of the most common malarial parasites, *Plasmodium falciparum*, by competitively inhibiting DHFR. While exhibiting a particularly low sequence conservation between species, most differences in sequence are from flexible loop regions [43], while the secondary structures between these loops are highly conserved across all species [44]. However, widespread use of pyrimethamine has resulted in a prevalence of pyrimethamine-resistant *P. falciparum* DHFR (pfDHFR) mutants as a result of four key amino acid substitutions at positions N51, C59, S108 and I164 which have also exhibited significant epistasis between mutation combinations [22] (Figure 5A).

An EpiScore analysis applied collectively to the behavior of the functionally important FG loop shows an immediate relationship between epistasis in pyrimethamine IC_{50} values of the pairwise mutants and their associated EpiScore values. Figure 5B shows the EpiScore violin plots (i.e., distributions and kernel density estimates) with respect to FG loop residues 196–206 for each pfDHFR mutant pair. These violin plots show that EpiScore distribution for different mutation pairs yields a different distribution for different residue pairs. S108-I164 gives a narrow distribution with a peak around 1, suggesting that force perturbations simultaneously exerted on these positions yields the same fluctuation response profile of the FG loop positions as the average of individual fluctuation responses of the FG loop when the forces are exerted individually at S108 and I164. This distribution pattern was also observed for positions N51 and C59, although at completely different positions within the protein. On the other hand, pairing the position I164 with C59 rather than with S108 results in a completely different EpiScore distribution, with diverse fluctuation responses of FG loop positions. This suggests that I164 and C59 are highly cooperative, leading to a non-additive behavior when these two positions are perturbed simultaneously. As I164 and C59 are located at different regions of the protein (Figure 5A), one can expect to observe a wide range of EpiScore values associated with this pair. This pattern tends to hold with distally located positions in the N51-S108 and C59-S108 distributions as well. Interestingly however, N51-C59 also exhibits EpiScore values less than 1, despite the fact that they belong to the same helical region. The distributions suggest that anisotropy in the network of interactions could modulate a wide range of fluctuation responses via these position pairs, which result in different functional behavior upon mutation. To determine whether the change in fluctuation response of the FG loop to simultaneous perturbations at these mutational positions can capture functional substitution outcomes, we next investigate the relationship between EpiScore and experimentally measured epistasis using pfDHFR pyrimethamine IC_{50} values.

Figure 5C presents the average EpiScore values with respect to the FG loop for each pfDHFR pairwise mutant, in order of increasing pyrimethamine IC_{50} epistasis. A dashed line at an EpiScore value of 1.0 has been added to aid in visual inspection. Here, IC_{50} epistasis is reported as the IC_{50} ratio of the dual mutant to the IC_{50} sum of the individual mutants. Any FG loop residue which was within 10 angstroms of either mutation site per dual mutant was excluded from the averaging in order to eliminate any strong dynamic coupling effects that arise as a result of direct contact interactions. The average EpiScore values have a strong, negative correlation ($R = -0.77$) with IC_{50} epistasis, where the stronger the positive epistasis, the lower the average EpiScore value. For example, an EpiScore value of ~0 means the pairwise dynamic coupling to FG loop positions of a dual mutant pair is negligible as compared to the average individual dynamic coupling; that is, the distal sites can individually impact position the FG loop residues allosterically. However, when treated together with a simultaneous perturbation at both sites, the displacement response of the FG loop residues are significantly lower, and, subsequently, their joint ability to allosterically regulate these FG loop positions is effectively lost. Due to the interaction network between the two distal positions with the FG loop, they may antagonistically compensate the amplitude and direction of the response when the perturbations on these two sites are exerted at the same time. To the reverse, an EpiScore value $\gg 1$ suggests that, simultaneously, two positions may exhibit dynamic coupling to the FG loop enough such that their pairwise mutational impact fundamentally alters the role the FG loop plays within the pfDHFR interaction network resulting in loss of function.

At first, this relationship may seem somewhat counterintuitive, as one could reasonably expect that the higher the EpiScore value (i.e., the stronger the dual position dynamic coupling versus individually averaged dynamic coupling), the higher the experimental epistasis. However, when complexed with substrate, the functionally critical M20 loop [45] is stabilized in part through interactions with amino acids in the FG loop [46]. It is possible that it is more favorable, in terms of pyrimethamine resistance, to have mutations occur at position pairs that induce a smaller fluctuation response of FG loop when perturbed simultaneously, (i.e., restricting the dynamics) than the average fluctuation response of individual perturbations applied one at a time. This is in agreement with previous work which showed that point mutations to two of the FG loop amino acids in E.coli resulted in a > 30 fold decrease in the steady state hydride transfer rate constant as compared to the wild-type [47]. This could additionally explain the pervasive and persistent nature of these mutations appearing globally in pfDHFR proteins.

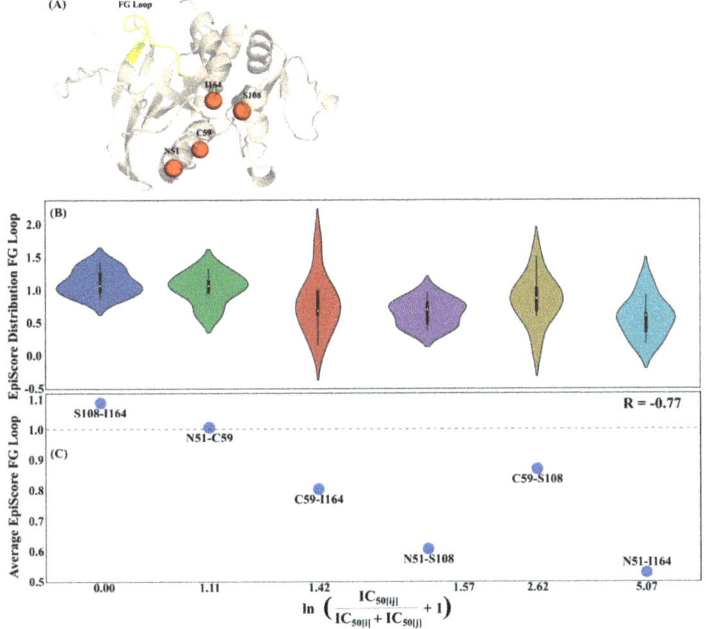

Figure 5. (**A**) *P. falciparum* DHFR (pfDHFR) structure (PDB ID 3QGT [48]) with FG loop residues (196–206) in yellow and mutation sites N51, C59, S108 and I164 colored in red. While not directly involved in catalytic activity, widespread use of pyrimethamine has resulted in pervasive and persistent mutations at these sites which confer pyrimethamine resistance. (**B**) Violin plot (distribution and kernel density estimate) of EpiScore values and (**C**) average EpiScore values with respect to FG loop residues for each pfDHFR dual mutant, in order of increasing pyrimethamine IC_{50} epistasis. A dashed line at EpiScore values of 1.0 has been added to aid in visual inspection. In (**B**) any FG loop residue which was within 10 angstroms of either mutation site per dual mutant was excluded from the averaging. The average EpiScore values have a strong, negative correlation (R = −0.77) with IC_{50} epistasis, where the stronger the positive epistasis, the lower the average EpiScore value.

Expanding our study to another system important to the concept of antibiotic resistance, we analyze TEM-1, a protein which possesses antibiotic resistance largely driven by its high evolvability, with over 170 TEM-1 mutants discovered as clinical or hospital isolates [49]. TEM-1 is a well-studied enzyme in experimental or laboratory-guided evolution, in an effort to both understand the mechanisms associated with its antibiotic resistance as well as predict possible resistance-conferring mutations [49–52].

Previous work has shown that the majority of the resistance-conferring mutations in TEM-1 are both distal to (10 Å or further) and highly coupled with the active site residues [53], indicating that these mutations impact TEM-1 function by allosterically regulating active site behavior. Additionally, it is now also understood that mutations resulting in the emergence of new enzymatic function are generally destabilizing which suggests that the evolution of new function requires additional, stabilizing mutations. As such, a more complete understanding of TEM-1 mutational behavior requires an investigation into the epistatic interplay of point mutation combinations. Thus, it is an ideal system for exploration of long-range dynamic communication to understand epistatic relationships in the emergence of resistance.

Here we focus on the specific epistatic relationship between four TEM-1 mutation sites (42, 104, 182 and 238) which have exhibited significant non-additive behavior [23]. Treating point mutations as external perturbative forces to the internal network of a protein, we apply EpiScore analysis to the main TEM-1 active site, residue S70, using a TEM-1 3-D structure obtained by an energy-minimized and equilibrated version of PDB ID 1BTL [53] with mutation sites shown as blue spheres in Figure 6A, along with active site S70 in red and alternative control sites (43, 105, 181 and 237) in yellow. Figure 6B (left) shows a plot of EpiScore versus experimental epistasis using cefotaxime turnover rates and exhibits a relationship similar to that found in pfDHFR, with a strong negative correlation of R = −0.71. We also find that position pairs with EpiScore values > 1.0 (horizontal dashed line), presenting a stronger pairwise dynamic coupling with position S70 compared to the average of the individual dynamic coupling, also corresponds to two of the three TEM-1 dual mutants with negative epistatic turnover rates (separated by vertical dashed line). Position pair 182-238 represents a deviation from this behavior, and position pair 42-104 is a comparative outlier to the overall correlation. The deviation of position pair 182-238 may be related to specific catalytic site interactions associated with position 238, the only position in which mutation resulted in an increase in turnover rate across all eight possible combinations of TEM-1 background. Interestingly, position 182, present in all position pairs with the three highest EpiScore values, was also the position in which mutation resulted in a significantly beneficial effect in the fewest number of possible backgrounds [23]. As a control, we also conducted this analysis using the alternative sites representing positions immediately adjacent to the four mutation sites (Figure 6B (right)). These positions result in a significantly worse correlation with cefotaxime turnover rate epistasis than the mutation positions (R = −0.45 as compared to R = −0.71), showing the sensitivity in the EpiScore metric to specific positions, regardless of separation distance.

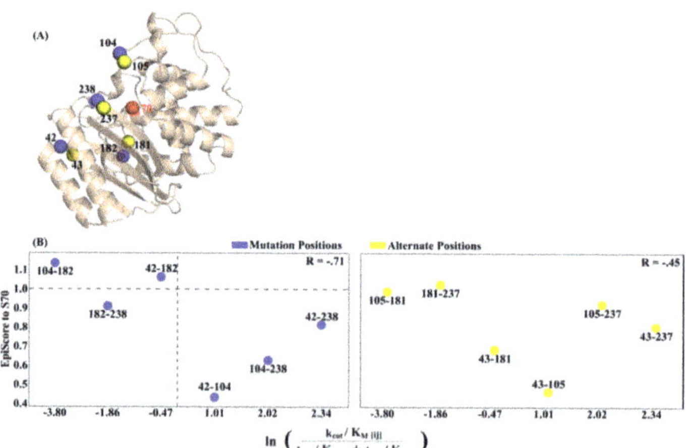

$$\ln \left(\frac{k_{cat}/K_{M\,[ij]}}{k_{cat}/K_{M\,[i]} + k_{cat}/K_{M\,[j]}} \right)$$

Figure 6. (**A**) TEM-1 structure showing mutation positions (blue spheres), alternative control positions (yellow spheres) and active site S70 (red sphere). ((**B**) left) EpiScore to active site S70 versus epistasis in ln of turnover rate of cefotaxime for β-lactamase TEM-1 mutants [23]. Horizontal dashed line divides EpiScore values above and below 1 while vertical dashed line divides positive and negative epistasis. EpiScore and epistasis exhibit a strong negative correlation of R = −0.71. EpiScore values > 1 (horizontal dashed line), indicating to a stronger pairwise dynamic coupling to position S70, also corresponds to two of the three TEM-1 dual mutants with negative epistatic turnover rates (separated by vertical dashed line). Position pair 182-238 represents a deviation from this behavior, and position pair 42-104 is a comparative outlier to the overall correlation. ((**B**) right) EpiScore versus ln of turnover rate using the alternative control positions. Although these positions are immediately adjacent to the mutation positions, they generate different EpiScore values resulting in a significantly worse correlation of R = −0.45.

3.2. Asymmetry and Epistasis

In TEM-1, mutational sites which confer incremental changes in biophysical activity are neither locally distributed with respect to one another, nor at important functional sites. Furthermore, they do not belong to an immediately identifiable allosteric inhibitor site, but they do, however, exhibit unique pairwise epistatic behavior which indicates that they likely regulate the active sites allosterically. In an effort to analyze whether the pairs having EpiScore less than 1 and associated with positive epistasis (a beneficial, cooperative interaction) exhibit long-range communication that is distinct from the pairs having EpiScore greater than 1 and associated with negative epistasis, we explored the degree of asymmetry in long-range communication between the mutational positions and the active site positions using DCI_{asym}. Thus, we calculated DCI_{asym} between each TEM-1 dual-mutant site and all main active sites for the relevant TEM-1 structure (70, 73, 130, 166, 234, Figure 7), excluding the outliers 182-238 and 42-104 from Figure 6B. Here, positive $\%DCI_{asym}$ values indicate active-site-dominant dynamic coupling, where mutational sites exhibit higher fluctuation response when the active site is perturbed. On the other hand, negative $\%DCI_{asym}$ values indicate mutation-dominant dynamic coupling where perturbations at those positions controls the active site fluctuation response. Interestingly, we observe a relationship that provides some mechanical insight relating the degree of asymmetry to EpiScore; the dual mutants with EpiScore > 1 to active site S70 and epistasis < 1 had more instances of mutational-dominant coupling asymmetry, while the reverse was true for two of the three dual mutants with EpiScore < 1 and epistasis > 1 (position pair 42-104, an outlier in Figure 6B, does not hold to this pattern). This suggests that the epistatic effects captured through EpiScore to active site S70 may be compensated via coupling asymmetry to all active sites, with dynamic modification of the system ultimately including both effects. A position pair that more strongly affects active site S70 via EpiScore also possesses active site-dominant coupling asymmetry and vice versa. Taken together with Figure 6B, these data indicate that dual mutants which confer less disruption to important active sites (indicated by EpiScore < 1) than their averaged individual constituents, and those which are under active site regulation, (indicated by positive $\%DCI_{asym}$) are those which display the largest degree of positive epistasis.

Thus, as a test system, TEM-1 highlights the complex relationship between mutational positions, allosteric relationships, and epistatic interplay. These emergent properties of the anisotropic residue-residue interaction network within a protein must be accounted for when attempting to fully understand or predict mutation outcomes.

3.3. Unidirectional Communication through DCI_{asym} Creates Cause-Effect Relationships in Allosteric Regulations

Using the dynamical picture presented above, the modulation of protein dynamics through mutations (i.e., the fluctuation response to node perturbations within a network) is similar to the modulation of dynamics through binding; this is the fundamental principle behind the concept of dynamic allostery. With the TEM-1 dual mutation positions showing unique coupling asymmetry to the active sites, it follows that there should be an obvious, unidirectional signature between allosteric sites and active sites in known allosteric proteins. Here we explore the role dynamic coupling directionality plays in allosteric regulations using an ideal model system, Pin1. Pin1 is a two-domain protein containing a catalytic PPIase domain and a distally-located WW domain, connected by a flexible (and highly disordered) interdomain linker [54–56]. While strictly regulated in both function and expression within healthy biological tissue [57], the up-regulation and down-regulation of Pin1 is associated with several forms of cancer and Alzheimer's disease, respectively [57–61]. Studies have shown that the activity of the PPIase domain is enhanced when a ligand is bound at the non-catalytic WW domain [62,63] and communication between these two domains is requisite for proper biological function [55,64–66].

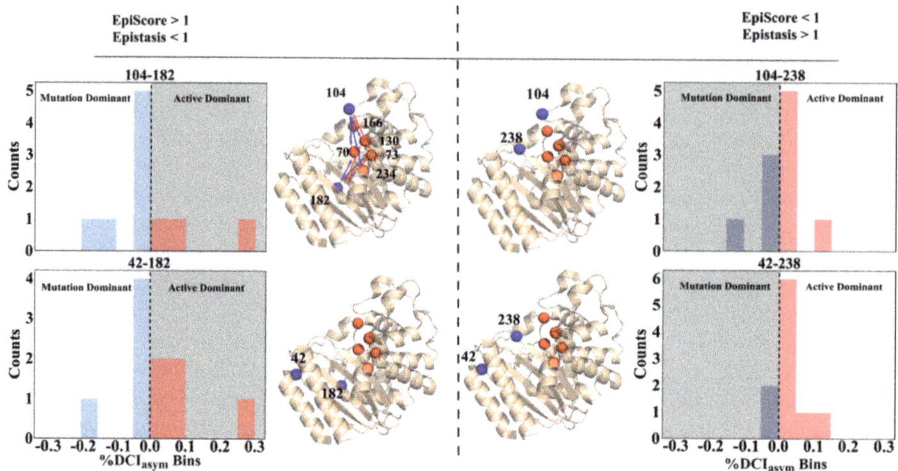

Figure 7. $\%DCI_{asym}$ distributions to all TEM-1 active sites (70, 73, 130, 166, 234) for each dual mutant position pair along with three-dimensional (3-D) structural representation of mutation sites (blue) and active sites (red) excluding the outliers 182-238 and 42-104 from Figure 6B. The first dual mutant (104-182, top left) has arrows drawn to indicate the coupling asymmetry, where red is active site-dominant and blue is mutation site-dominant. Both dual mutants with negative epistasis in turnover rate and EpiScore to position S70 < 1 also had $\%DCI_{asym}$ distributions which were, overall, mutation site-dominant and, conversely, those with positive epistasis in turnover rate and EpiScore to position S70 > 1 exhibited active site-dominant $\%DCI_{asym}$.

Previous works propose the existence of communication networks between the WW domain and the PPIase domain, including a unique allosteric pathway which only becomes active when a substrate is bound to the WW domain [62]. A further computational study indicated that pathways of communication via force propagation from the PPIase domain to the WW domain changed when a ligand was WW domain-bound [67].

Applying our asymmetry analysis to binding pocket residues in the catalytic PPIase ($\%DCI_{ij}$) and non-catalytic WW domains ($\%DCI_{ji}$) of Pin1 (PDB ID 3TCZ [7], ligands removed), we calculate "$\%DCI_{asym}$" ($\%DCI_{ij} - \%DCI_{ji}$) the coupling asymmetry between PPIase domain binding positions (63, 68, 129, 130, 131, 154) and WW domain binding positions (23, 31, 32, 34) (Figure 8). Hence, negative values indicate the WW domain position is dominant (blue arrows) whereas positive values indicate the PPIase domain position is dominant (red arrows). We see that each of the four positions in the WW domain exhibit unique asymmetric coupling with the PPIase domain positions, even when the WW domain positions are close to one another. However, with the exception of coupling between position 63 and 31, the behavior of the PPIase domain positions is unique to their catalytic loop grouping (e.g., {63,68}, {129,130,131}), where each position within a group has the same asymmetry directionality to a given WW domain position. Overall, however, the full $\%DCI_{asym}$ distribution indicates that there is a clear bias toward unidirectionality from the WW domain to the PPIase domain; the WW domain is dynamic coupling-dominant over the PPIase domain, with twice as many residue pairs exhibiting WW-dominant coupling than the reverse (16/24 vs. 8/24, Figure 8C).

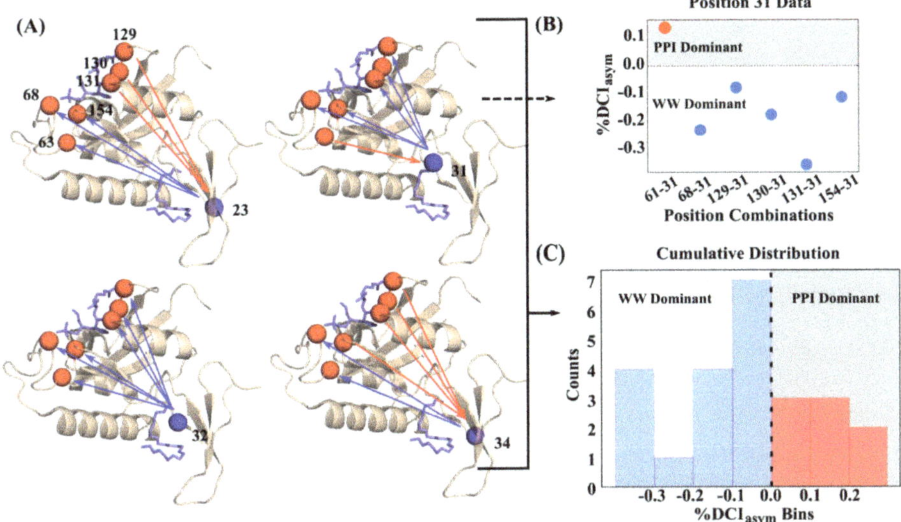

Figure 8. (**A**) Graphical representation of coupling asymmetry between PPIase domain binding positions (63, 68, 129, 130, 131, 154) and WW domain binding positions (23, 31, 32, 34) where blue arrows indicate the WW domain position is dominant and red arrows indicate the PPIase domain position is dominant. (**B**) Explicit values of $\%DCI_{asym}$ versus position combinations for position 31 in the WW binding domain where values above and below 0 correspond to PPIase or WW domain dominating, respectively (a value of 0 corresponds to perfect symmetry). (**C**) Full distribution of $\%DCI_{asym}$ values for all four WW domain binding positions and all six PPIase domain binding positions where 16 total residue pairs are dominated by the WW domain whereas only eight pairs are PPIase domain-dominant.

This suggests a cause-and-effect relationship exists between the two domains. Using this framework, a ligand binding event is modeled as a force perturbation to the binding positions in each domain. Upon these random force perturbations, we find that, overall, the WW domain is able to induce a stronger perturbation response in the PPIase domain than the reverse. This is largely the expected relationship between an allosteric site and a catalytic site; communication between these sites should predominantly involve information transfer from the allosteric site to the catalytic site, indicating that $\%DCI_{asym}$ can capture communication directionality in allosteric systems from structural dynamics encoded within a given set of atomic coordinates.

4. Conclusions

In this work we showed how the anisotropic interaction network within a protein captures two essential emergent properties of protein evolution—epistasis and communication directionality—using the information stored in structural dynamics alone. Additionally, EpiScore can capture the behavior of dual-mutation epistatic outcomes with some consistent trends across different protein systems. As seen in pfDHFR, mutation pairs with a lower pairwise dynamic coupling versus average of individual couplings (EpiScore < 1) to FG loop positions are favorable, as dual mutations at these positions may be less likely to disrupt the FG loop's interaction with the functionally critical M20 loop. A similar trend was also observed in the EpiScore analysis of TEM-1 dual mutants, where lower EpiScore to active site S70 was generally associated with higher positive experimental epistasis (R = −0.71) Further, the system-wide EpiScore analysis of GB1 dual mutants has shown that the position pairs with average EpiScore values > 1 were associated more frequently with negative epistasis, indicating that these positions might ultimately be more disruptive to the entire protein when mutated together. Furthermore, when dynamic coupling asymmetry analysis was applied via $\%DCI_{asym}$ to TEM-1, we found that EpiScore and epistasis both relate to dynamic coupling asymmetry, where position pairs which exhibited high EpiScores associated with negative epistasis also exhibited mutation-dominant coupling asymmetry. This suggests that $\%DCI_{asym}$ and EpiScore may both capture factors which contribute towards the biochemical outcome of dual mutations. If both mutational sites dominate the dynamics coupling with the active site (i.e., the active site responds more to mutational site perturbations), then dual mutations on both sites lead to negative epistasis.

As modulation of normal modes and protein dynamics is not only a tool used in evolution but also a principle exploited via allostery, we used an "ideal" allosteric system, Pin1, and observed the dynamic coupling asymmetry between a well-identified allosteric domain and an enzymatically active domain exhibits behavior that, as expected, showed the allosteric WW domain to dominate communication to the PPIase domain. Overall, these two novel protein dynamics-based metrics provide steps to mechanistically describe these complicated interactions, and also shed light on the complex anisotropic interaction network which ultimately gives rise to epistasis and allosteric regulation. They can be useful to predict mutational outcomes, particularly for those sites distal from the active site that can modulate function [68].

Author Contributions: P.C. and S.B.O. constructed the theory and methodology. P.C. performed the analysis and both P.C. and S.B.O. wrote the manuscript. All authors have read and agreed to the published version of the manuscript.

Funding: This work is supported by the National Science Foundation Division of Molecular and Cellular Biosciences (award 1715591) and the Gordan and Betty Moore Foundation.

Conflicts of Interest: The authors declare no conflict of interest.

References

1. Subramanian, S.; Kumar, S. Evolutionary anatomies of positions and types of disease-associated and neutral amino acid mutations in the human genome. *BMC Genom.* **2006**, *7*, 306. [CrossRef] [PubMed]
2. Kumar, S.; Dudley, J.T.; Filipski, A.; Liu, L. Phylomedicine: An evolutionary telescope to explore and diagnose the universe of disease mutations. *Trends Genet.* **2011**, *27*, 377–386. [CrossRef]
3. Swint-Kruse, L. Using evolution to guide protein engineering: The devil is in the details. *Biophys. J.* **2016**, *111*, 10–18. [CrossRef]
4. Starr, T.N.; Thornton, J.W. Epistasis in protein evolution. *Protein Sci.* **2016**, *25*, 1204–1218. [CrossRef]
5. Domingo, J.; Diss, G.; Lehner, B. Pairwise and higher-order genetic interactions during the evolution of a tRNA. *Nature* **2018**, *558*, 117–121. [CrossRef]
6. Salinas, V.H.; Ranganathan, R. Coevolution-based inference of amino acid interactions underlying protein function. *Elife* **2018**, *7*, e34300. [CrossRef]
7. Zhang, M.; Wang, X.J.; Chen, X.; Bowman, M.E.; Luo, Y.; Noel, J.P.; Ellington, A.D.; Etzkorn, F.A.; Zhang, Y. Structural and kinetic analysis of prolyl-isomerization/phosphorylation cross-talk in the CTD code. *ACS Chem. Biol.* **2012**, *7*, 1462–1470. [CrossRef]
8. Figliuzzi, M.; Jacquier, H.; Schug, A.; Tenaillon, O.; Weigt, M. Coevolutionary landscape inference and the context-dependence of mutations in beta-lactamase TEM-1. *Mol. Biol. Evol.* **2016**, *33*, 268–280. [CrossRef]
9. Wang, Z.O.; Pollock, D.D. Context dependence and coevolution among amino acid residues in proteins. *Methods Enzymol.* **2005**, *395*, 779–790. [CrossRef]
10. Payne, J.L.; Wagner, A. The causes of evolvability and their evolution. *Nat. Rev. Genet.* **2019**, *20*, 24–38. [CrossRef] [PubMed]
11. Larrimore, K.E.; Kazan, I.C.; Kannan, L.; Kendle, R.P.; Jamal, T.; Barcus, M.; Bolia, A.; Brimijoin, S.; Zhan, C.-G.; Ozkan, S.B.; et al. Plant-expressed cocaine hydrolase variants of butyrylcholinesterase exhibit altered allosteric effects of cholinesterase activity and increased inhibitor sensitivity. *Sci. Rep.* **2017**, *7*, 10419. [CrossRef]
12. Kumar, A.; Glembo, T.J.; Ozkan, S.B. The role of conformational dynamics and allostery in the disease development of human ferritin. *Biophys. J.* **2015**, *109*, 1273–1281. [CrossRef]

13. Modi, T.; Huihui, J.; Ghosh, K.; Ozkan, S.B. Ancient thioredoxins evolved to modern-day stability–function requirement by altering native state ensemble. *Philos. Trans. R. Soc. Lond. B Biol. Sci.* **2018**, *373*, 20170184. [CrossRef] [PubMed]

14. Tokuriki, N.; Tawfik, D.S. Protein dynamism and evolvability. *Science* **2009**, *324*, 203–207. [CrossRef]

15. Bhabha, G.; Ekiert, D.C.; Jennewein, M.; Zmasek, C.M.; Tuttle, L.M.; Kroon, G.; Dyson, H.J.; Godzik, A.; Wilson, I.A.; Wright, P.E. Divergent evolution of protein conformational dynamics in dihydrofolate reductase. *Nat. Struct. Mol. Biol.* **2013**, *20*, 1243–1249. [CrossRef]

16. Meinhardt, S.; Manley, M.W.; Parente, D.J.; Swint-Kruse, L. Rheostats and toggle switches for modulating protein function. *PLoS ONE* **2013**, *8*, e83502. [CrossRef]

17. Saavedra, H.G.; Wrabl, J.O.; Anderson, J.A.; Li, J.; Hilser, V.J. Dynamic allostery can drive cold adaptation in enzymes. *Nature* **2018**, *558*, 324–328. [CrossRef] [PubMed]

18. McLeish, T.C.B.; Rodgers, T.L.; Wilson, M.R. Allostery without conformation change: Modelling protein dynamics at multiple scales. *Phys. Biol.* **2013**, *10*, 56004. [CrossRef]

19. Gobeil, S.M.C.; Ebert, M.C.C.J.C.; Park, J.; Gagné, D.; Doucet, N.; Berghuis, A.M.; Pleiss, J.; Pelletier, J.N. The structural dynamics of engineered β-lactamases vary broadly on three timescales yet sustain native function. *Sci. Rep.* **2019**, *9*, 6656. [CrossRef]

20. Rollins, N.J.; Brock, K.P.; Poelwijk, F.J.; Stiffler, M.A.; Gauthier, N.P.; Sander, C.; Marks, D.S. Inferring protein 3D structure from deep mutation scans. *Nat. Genet.* **2019**, *51*, 1170–1176. [CrossRef]

21. Faber, M.S.; Wrenbeck, E.E.; Azouz, L.R.; Steiner, P.J.; Whitehead, T.A. Impact of In Vivo Protein Folding Probability on Local Fitness Landscapes. *Mol. Biol. Evol.* **2019**, *36*, 2764–2777. [CrossRef] [PubMed]

22. Lozovsky, E.R.; Chookajorn, T.; Brown, K.M.; Imwong, M.; Shaw, P.J.; Kamchonwongpaisan, S.; Neafsey, D.E.; Weinreich, D.M.; Hartl, D.L. Stepwise acquisition of pyrimethamine resistance in the malaria parasite. *Proc. Natl. Acad. Sci. USA* **2009**, *106*, 12025–12030. [CrossRef] [PubMed]

23. Knies, J.L.; Cai, F.; Weinreich, D.M. Enzyme efficiency but not thermostability drives cefotaxime resistance evolution in TEM-1 β-lactamase. *Mol. Biol. Evol.* **2017**, *34*, 1040–1054. [CrossRef] [PubMed]

24. Olson, C.A.; Wu, N.C.; Sun, R. A comprehensive biophysical description of pairwise epistasis throughout an entire protein domain. *Curr. Biol.* **2014**, *24*, 2643–2651. [CrossRef]

25. Varona, L.; Legarra, A.; Toro, M.A.; Vitezica, Z.G. Non-additive effects in genomic selection. *Front. Genet.* **2018**, *9*, 78. [CrossRef]

26. Nussinov, R.; Tsai, C.-J. Allostery in disease and in drug discovery. *Cell* **2013**, *153*, 293–305. [CrossRef] [PubMed]

27. Nussinov, R.; Tsai, C.-J.; Liu, J. Principles of allosteric interactions in cell signaling. *J. Am. Chem. Soc.* **2014**, *136*, 17692–17701. [CrossRef]

28. Popovych, N.; Sun, S.; Ebright, R.H.; Kalodimos, C.G. Dynamically driven protein allostery. *Nat. Struct. Mol. Biol.* **2006**, *13*, 831–838. [CrossRef]

29. Tsai, C.-J.; del Sol, A.; Nussinov, R. Allostery: Absence of a change in shape does not imply that allostery is not at play. *J. Mol. Biol.* **2008**, *378*, 1–11. [CrossRef]

30. Zhang, Y.; Doruker, P.; Kaynak, B.; Zhang, S.; Krieger, J.; Li, H.; Bahar, I. Intrinsic dynamics is evolutionarily optimized to enable allosteric behavior. *Curr. Opin. Struct. Biol.* **2019**, *62*, 14–21. [CrossRef]

31. Loutchko, D.; Flechsig, H. Allosteric communication in molecular machines via information exchange: What can be learned from dynamical modeling. *Biophys. Rev.* **2020**, *12*, 443–452. [CrossRef]

32. Dutta, S.; Eckmann, J.-P.; Libchaber, A.; Tlusty, T. Green function of correlated genes in a minimal mechanical model of protein evolution. *Proc. Natl. Acad. Sci. USA* **2018**, *115*, E4559–E4568. [CrossRef]

33. Flechsig, H.; Togashi, Y. Designed Elastic Networks: Models of Complex Protein Machinery. *Int. J. Mol. Sci.* **2018**, *19*, 3152. [CrossRef] [PubMed]

34. Bandaru, P.; Shah, N.H.; Bhattacharyya, M.; Barton, J.P.; Kondo, Y.; Cofsky, J.C.; Gee, C.L.; Chakraborty, A.K.; Kortemme, T.; Ranganathan, R.; et al. Deconstruction of the Ras switching cycle through saturation mutagenesis. *Elife* **2017**, *6*, e27810. [CrossRef] [PubMed]

35. Fowler, D.M.; Fields, S. Deep mutational scanning: A new style of protein science. *Nat. Methods* **2014**, *11*, 801–807. [CrossRef] [PubMed]

36. Romero, P.A.; Tran, T.M.; Abate, A.R. Dissecting enzyme function with microfluidic-based deep mutational scanning. *Proc. Natl. Acad. Sci. USA* **2015**, *112*, 7159–7164. [CrossRef]

37. Melamed, D.; Young, D.L.; Gamble, C.E.; Miller, C.R.; Fields, S. Deep mutational scanning of an RRM domain of the Saccharomyces cerevisiae poly(A)-binding protein. *RNA* **2013**, *19*, 1537–1551. [CrossRef]

38. Gerek, Z.N.; Ozkan, S.B. Change in allosteric network affects binding affinities of PDZ domains: Analysis through perturbation response scanning. *PLoS Comput. Biol.* **2011**, *7*, e1002154. [CrossRef] [PubMed]

39. Nevin Gerek, Z.; Kumar, S.; Banu Ozkan, S. Structural dynamics flexibility informs function and evolution at a proteome scale. *Evol. Appl.* **2013**, *6*, 423–433. [CrossRef]

40. Schrodinger L.L.C. *The PyMOL Molecular Graphics System.* Available online: https://pymol.org/2/ (accessed on 10 June 2020).

41. Jelsch, C.; Mourey, L.; Masson, J.M.; Samama, J.P. Crystal structure of Escherichia coli TEM1 beta-lactamase at 1.8 A resolution. *Proteins* **1993**, *16*, 364–383. [CrossRef] [PubMed]

42. Frericks Schmidt, H.L.; Sperling, L.J.; Gao, Y.G.; Wylie, B.J.; Boettcher, J.M.; Wilson, S.R.; Rienstra, C.M. Crystal polymorphism of protein GB1 examined by solid-state NMR and X-ray diffraction. *J. Phys. Chem. B* **2007**, *111*, 14362–14369. [CrossRef] [PubMed]

43. Liu, C.T.; Hanoian, P.; French, J.B.; Pringle, T.H.; Hammes-Schiffer, S.; Benkovic, S.J. Functional significance of evolving protein sequence in dihydrofolate reductase from bacteria to humans. *Proc. Natl. Acad. Sci. USA* **2013**, *110*, 10159–10164. [CrossRef] [PubMed]

44. Cody, V.; Schwalbe, C.H. Structural characteristics of antifolate dihydrofolate reductase enzyme interactions. *Crystallogr. Rev.* **2006**, *12*, 301–333. [CrossRef]

45. Mhashal, A.R.; Vardi-Kilshtain, A.; Kohen, A.; Major, D.T. The role of the Met20 loop in the hydride transfer in Escherichia coli dihydrofolate reductase. *J. Biol. Chem.* **2017**, *292*, 14229–14239. [CrossRef] [PubMed]

46. Sawaya, M.R.; Kraut, J. Loop and subdomain movements in the mechanism of Escherichia coli dihydrofolate reductase: Crystallographic evidence. *Biochemistry* **1997**, *36*, 586–603. [CrossRef]

47. Behiry, E.M.; Evans, R.M.; Guo, J.; Loveridge, E.J.; Allemann, R.K. Loop interactions during catalysis by dihydrofolate reductase from Moritella profunda. *Biochemistry* **2014**, *53*, 4769–4774. [CrossRef]

48. Vanichtanankul, J.; Taweechai, S.; Yuvaniyama, J.; Vilaivan, T.; Chitnumsub, P.; Kamchonwongpaisan, S.; Yuthavong, Y. Trypanosomal dihydrofolate reductase reveals natural antifolate resistance. *ACS Chem. Biol.* **2011**, *6*, 905–911. [CrossRef] [PubMed]

49. Salverda, M.L.M.; de Visser, J.A.G.M.; Barlow, M. Natural evolution of TEM-1 β-lactamase: Experimental reconstruction and clinical relevance. *FEMS Microbiol. Rev.* **2010**, *34*, 1015–1036. [CrossRef]

50. Ingles-Prieto, A.; Ibarra-Molero, B.; Delgado-Delgado, A.; Perez-Jimenez, R.; Fernandez, J.M.; Gaucher, E.A.; Sanchez-Ruiz, J.M.; Gavira, J.A. Conservation of protein structure over four billion years. *Structure* **2013**, *21*, 1690–1697. [CrossRef]

51. Weinreich, D.M.; Delaney, N.F.; Depristo, M.A.; Hartl, D.L. Darwinian evolution can follow only very few mutational paths to fitter proteins. *Science* **2006**, *312*, 111–114. [CrossRef]

52. Zou, T.; Risso, V.A.; Gavira, J.A.; Sanchez-Ruiz, J.M.; Ozkan, S.B. Evolution of conformational dynamics determines the conversion of a promiscuous generalist into a specialist enzyme. *Mol. Biol. Evol.* **2015**, *32*, 132–143. [CrossRef]

53. Modi, T.; Ozkan, B. Allostery modulates resistance driver mutations in TEM-1. *Biophys. J.* **2019**, *116*, 342a. [CrossRef]

54. Zhang, Y.; Daum, S.; Wildemann, D.; Zhou, X.Z.; Verdecia, M.A.; Bowman, M.E.; Lücke, C.; Hunter, T.; Lu, K.-P.; Fischer, G.; et al. Structural basis for high-affinity peptide inhibition of human Pin1. *ACS Chem. Biol.* **2007**, *2*, 320–328. [CrossRef] [PubMed]

55. Verdecia, M.A.; Bowman, M.E.; Lu, K.P.; Hunter, T.; Noel, J.P. Structural basis for phosphoserine-proline recognition by group IV WW domains. *Nat. Struct. Biol.* **2000**, *7*, 639–643. [CrossRef] [PubMed]

56. Ranganathan, R.; Lu, K.P.; Hunter, T.; Noel, J.P. Structural and functional analysis of the mitotic rotamase Pin1 suggests substrate recognition is phosphorylation dependent. *Cell* **1997**, *89*, 875–886. [CrossRef]

57. Lu, K.P.; Zhou, X.Z. The prolyl isomerase PIN1: A pivotal new twist in phosphorylation signalling and disease. *Nat. Rev. Mol. Cell Biol.* **2007**, *8*, 904–916. [CrossRef] [PubMed]

58. Balastik, M.; Lim, J.; Pastorino, L.; Lu, K.P. Pin1 in Alzheimer's disease: Multiple substrates, one regulatory mechanism? *Biochim. Biophys. Acta-Mol. Bas. Dis.* **2007**, *1772*, 422–429. [CrossRef]

59. Lu, K.P. Pinning down cell signaling, cancer and Alzheimer's disease. *Trends Biochem. Sci.* **2004**, *29*, 200–209. [CrossRef]

60. Wulf, G.M.; Ryo, A.; Wulf, G.G.; Lee, S.W.; Niu, T.; Petkova, V.; Lu, K.P. Pin1 is overexpressed in breast cancer and cooperates with Ras signaling in increasing the transcriptional activity of c-Jun towards cyclin D1. *EMBO J.* **2001**, *20*, 3459–3472. [CrossRef] [PubMed]

61. Zhou, X.Z.; Lu, K.P. The isomerase PIN1 controls numerous cancer-driving pathways and is a unique drug target. *Nat. Rev. Cancer* **2016**, *16*, 463–478. [CrossRef]

62. Guo, J.; Pang, X.; Zhou, H.-X. Two pathways mediate inter-domain allosteric regulation in Pin1. *Structure* **2014**, *23*, 237–247. [CrossRef] [PubMed]

63. Peng, J.W. Investigating dynamic interdomain allostery in Pin1. *Biophys. Rev.* **2015**, *7*, 239–249. [CrossRef] [PubMed]

64. Li, Z.; Li, H.; Devasahayam, G.; Gemmill, T.; Chaturvedi, V.; Hanes, S.D.; van Roey, P. The structure of the Candida albicans Ess1 prolyl isomerase reveals a well-ordered linker that restricts domain mobility. *Biochemistry* **2005**, *44*, 6180–6189. [CrossRef] [PubMed]

65. Namanja, A.T.; Wang, X.J.; Xu, B.; Mercedes-Camacho, A.Y.; Wilson, K.A.; Etzkorn, F.A.; Peng, J.W. Stereospecific gating of functional motions in Pin1. *Proc. Natl. Acad. Sci. USA* **2011**, *108*, 12289–12294. [CrossRef]

66. Peng, J.W.; Wilson, B.D.; Namanja, A.T. Mapping the dynamics of ligand reorganization via 13CH3 and 13CH2 relaxation dispersion at natural abundance. *J. Biomol. NMR* **2009**, *45*, 171–183. [CrossRef]

67. Campitelli, P.; Guo, J.; Zhou, H.-X.; Ozkan, S.B. Hinge-shift mechanism modulates allosteric regulations in human Pin1. *J. Phys. Chem. B* **2018**, *122*, 5623–5629. [CrossRef]

68. Campitelli, P.; Modi, T.; Kumar, S.; Ozkan, S.B. The role of conformational dynamics and allostery in modulating protein evolution. *Annu. Rev. Biophys.* **2020**, *49*, 267–288. [CrossRef]

entropy

Article

Dynamical Behavior of β-Lactamases and Penicillin-Binding Proteins in Different Functional States and Its Potential Role in Evolution

Feng Wang [1], Hongyu Zhou [1], Xinlei Wang [2] and Peng Tao [1,*]

[1] Department of Chemistry, Center for Drug Discovery, Design, and Delivery (CD4), Center for Scientific Computation, Southern Methodist University, Dallas, TX 75275, USA; fengw@smu.edu (F.W.); hongyuz@smu.edu (H.Z.)

[2] Department of Statistical Science, Southern Methodist University, Dallas, TX 75275, USA; swang@smu.edu

* Correspondence: ptao@smu.edu

Received: 15 September 2019; Accepted: 15 November 2019; Published: 19 November 2019

Abstract: β-Lactamases are enzymes produced by bacteria to hydrolyze β-lactam-based antibiotics, and pose serious threat to public health through related antibiotic resistance. Class A β-lactamases are structurally and functionally related to penicillin-binding proteins (PBPs). Despite the extensive studies of the structures, catalytic mechanisms and dynamics of both β-lactamases and PBPs, the potentially different dynamical behaviors of these proteins in different functional states still remain elusive in general. In this study, four evolutionarily related proteins, including TEM-1 and TOHO-1 as class A β-lactamases, PBP-A and DD-transpeptidase as two PBPs, are subjected to molecular dynamics simulations and various analyses to characterize their dynamical behaviors in different functional states. Penicillin G and its ring opening product serve as common ligands for these four proteins of interest. The dynamic analyses of overall structures, the active sites with penicillin G, and three catalytically important residues commonly shared by all four proteins reveal unexpected cross similarities between Class A β-lactamases and PBPs. These findings shed light on both the hidden relations among dynamical behaviors of these proteins and the functional and evolutionary relations among class A β-lactamases and PBPs.

Keywords: TEM-1; TOHO-1; PBP-A; DD-transpeptidase; conformational changes; catalytic mechanism; evolution

---

1. Introduction

β-Lactam antibiotics have been used to treat bacterial infections since 1942. Antibiotics can interfere with the cross-linking in cell-wall biosynthesis, inhibiting cell wall growth and thus killing bacteria. As a mechanism of resistance for survival, bacteria produce β-lactamases to inactivate β-lactam antibiotics. The bacterial resistance to β-lactam antibiotic is an urgent and critical threat to global health. The main resistance mechanism involves antibiotic hydrolysis by β-lactamases through acylation and de-acylation catalytic cycles. β-Lactamases can be classified into four sub-groups (Classes A, B, C and D) based on their amino acid sequences and substrates [1]. Classes A, C and D are serine-based β-lactamases and class B are zinc-based β-lactamases. Among the four sub-groups, class A β-lactamases have a wide range of substrates and can spread via horizontal transfer, therefore posing a serious threat to public health [2] and have been widely studied [3].

Penicillin is the first commonly used β-lactam antibiotic. TEM-1, belonging to the Class A β-lactamases, can hydrolyze penicillin with high efficiency. TEM-1 has two conserved domains (α/β and α) around its active site. The structure of TEM-1 binding with benzyl penicillin (penicillin G) as substrate (PDB ID: 1FQG) at 1.7 Å resolution was reported in 1992 [4]. In the reported crystal structure, the penicillin G has a covalent bond with Ser70 as an intermediate structure. The catalytic mechanism of TEM-1 against penicillin was proposed to involve acylation and de-acylation steps (Figure 1). In the acylation step, the β-lactam ring of penicillin is attacked by the TEM-1 Ser70 residue. In the de-acylation step, TEM-1 Glu166 residue acts as a general base in the attack on the substrate assisted by a water molecule.

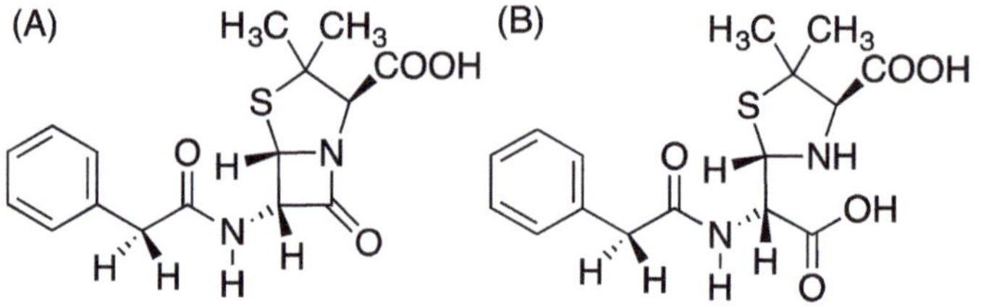

Figure 1. (**A**) Benzyl penicillin, (**B**) the hydrolysis product of benzyl penicillin.

TOHO-1 is another Class A β-lactamase, and has an efficient hydrolytic activity against penicillin. Among Class A β-lactamases, TEM and the extended spectrum β-lactamases (ESBLs) CTX-M exhibit highly variable substrate profiles. TOHO-1 belongs to the ESBL CTX-M type enzymes and exhibits about 40% identity with the TEM families. The acyl-intermediate structure of TOHO-1 with penicillin G was reported in 2002 [5]. Like other β-lactamases, TOHO-1 has two highly conserved domains (α/β and α) around its active site. Ser70 at the active site of TOHO-1 is also critical for the hydrolysis of the penicillin molecule. The structure TOHO-1 apo forms with triple mutants was solved using neutron diffraction [6]. The Glu166 residue was proposed to act as a general base in the acylation reactions of TOHO-1 [7]. The catalytic mechanism of TOHO-1 against cefotaxime was investigated using both neutron and high-resolution X-ray diffraction. The study further emphasized the role of Lys73 in the acylation mechanism [8]. Additional studies also related TOHO-1 catalytic mechanism with the functions of active site residues [9–11].

Many reported structural evidences show that β-lactamases were evolved from cell wall biosynthetic enzymes, which are referred to as penicillin-binding proteins (PBPs) [12]. PBPs have a high sequence homology to class A β-lactamases. For example, the protein PBP-A shares a typical catalytic cavity and an overall fold with Class A β-lactamases, and 28% sequence similarities with β-lactamases on average [13]. Comparing the structure of PBP-A with Class A β-lactamases, PBP-A has a six residues deletion on the conserved Ω-loop, and there is no residue corresponding to Glu166 in TEM-1 as a general base in hydrolysis mechanism.

D-Alanyl-D-alanine transpeptidase (DD-transpeptidase), which was discovered in *Streptomyces* sp. R61 and classified as a PBP, has low sequence similarity compared to class A β-lactamases. DD-transpeptidase is a main target of penicillin and was proposed to share the same ancestor with β-lactamases [14]. Similar to TEM-1 and TOHO-1, DD-transpeptidase also has acylation and de-acylation catalytic steps for hydrolyzing penicillin. One main difference between β-lactamases (TEM-1 and TOHO-1) and DD-transpeptidase is that the de-acylation step reaction rate of DD-transpeptidase is extremely slow. Therefore, DD-transpeptidase can become effectively trapped in the acylated state. The crystal structure of DD-transpeptidase with penicillin G as a substrate was solved in 2004 and used to compare with the DD-transpeptidase complex with a peptidoglycan-mimetic β-lactam [15]. The structures and kinetic data from this study support the hypothesis that peptidoglycan-mimetic side-chains can improve the β-lactam inhibition activity [15]. In addition, Thr299 was identified as a highly conserved residue in the active site of DD-transpeptidase [16].

It was proposed that the majority of Class A β-lactamases and the present DD-transpeptidase were evolved from a same ancestor, most likely a DD-transpeptidase because of the similarity of substrate profiles, overall folds and the functional groups of the active site [17]. The ligand similarity could be used to cluster proteins with low sequence similarity. For example, a network-based model reported by Cheng et al. analyzes the interaction between proteins and ligands without structural information [18]. β-Lactamases and PBPs haven been studied through a ligand centric network model as well [19].

PBPs could undergo acylation reaction with penicillin G. PBP-A was identified as a member of a new family in PBPs due to a significant sequence similarity to class A β-lactamases. The crystal structure of PBP-A in apo state and acylated with penicillin G intermediate are both available from a study to evolve PBP-A into β-lactamase. In this study, PBP-A was compared with TEM-1 using structural alignment and hydrogen bond networks analysis [13]. Residue Glu166 was introduced to the shorter Ω-Loop of PBP-A, and a 90-fold increase in de-acylation rate was obtained. However, the sequence of PBP-A was not homologous with DD-transpeptidase [20].

Many computational methods, including molecular dynamic (MD) simulations and hybrid quantum mechanical and molecular mechanical (QM/MM) calculations were used to characterize the conformational changes and elucidate the catalytic mechanisms of protein structures [21]. The catalytic mechanism of DD-transpeptidase against cephalothin was studied using QM/MM method [22]. Tyr159 in DD-transpeptidase was proposed to carry functions different from the Tyr150 in β-lactamase as a general base. β-Lactamases have a rapid de-acylation step compared with DD-transpeptidase [23]. MD simulations could provide detailed dynamical insight into protein functions. Markov state model (MSM) is an effective method to model the kinetic information based on MD simulations [24,25].

In one of our previous studies, the dynamical properties of TEM-1 in different functional states, including complexes binding with penicillin G and its de-acylation product, respectively, and the apo state were characterized through MD simulations and machine learning methods [26]. The key residues for TEM-1 dynamics in different functional states were identified using machine learning methods.

In the current study, TEM-1, TOHO-1, PBP-A and DD-transpeptidase as structurally and functionally related proteins are subjected to extensive MD simulations and analyses. Despite the low sequence and structural homology, all four proteins could hydrolyze penicillin G, and are coupled with the evolution from ancient PBPs to β-lactamases. Detailed analyses were carried out using the hidden Markov state model (HMM) based on the overall enzyme structures and MSMs based on the active site binding with penicillin G to shed light onto the evolutionary relations among these four proteins.

2. Materials and Methods

The initial structures of TEM-1, TOHO-1, Penicillin-Binding Protein (PBP-A) and DD-transpeptidase were obtained from the Protein Data Bank (PDB). Their PDB IDs are 1FQG [4], 1IYQ [5], 2J8Y [13] and 1PWC [15], respectively. All four selected crystal structures are in acylated state with covalent bonds to penicillin G intermediates. For each protein, three states were constructed: the apo state (a protein alone without ligand), reactant state (a protein binding with penicillin G, Figure 2A), and product state (a protein binding with the hydrolyzed product of penicillin G Figure 2B). It should be noted that no crystal structure for these enzymes in the reactant or product state is available in PDB.

2.1. Molecular Dynamic Simulations

The CHARMM36 force field was used to describe the selected enzymes [27]. CHARMM General Force Fields (CGenFF) [28,29] for penicillin G and its de-acylation product were generated using the online server ParamChem (https://cgenff.paramchem.org/). All simulation systems were solvated in water box using a TIP3P water model [30,31] with the addition of sodium and chloride ions to balance the charge and reproduce typical physiological ion concentrations. Initially, the simulation systems were subjected to 5000 steps of the steepest descent energy minimization and the adopted basis Newton-Raphson (ABNR) method with the gradient tolerance 0.02 kcal/mol·Å. Then, 10 ns of isothermal-isobaric ensemble (NPT) MD simulation was carried out for four proteins in each state. Subsequently, 1050 ns NVT ensemble MD simulations at 300 K were conducted. The first 50 ns simulations were discarded as equilibration and the following 1 μs was used for further analysis. The time step for MD simulations is 2 fs, and simulation trajectories were saved every 1000 steps (2 ps). All the bonds associated with hydrogen atoms were fixed during the simulation using SHAKE method [32]. Periodic boundary condition was used in all simulations, and electrostatic interactions were calculated using the particle mesh Ewald (PME) method [33]. All structural preparation and simulations were constructed using CHARMM simulation package version 41b1 with the support of GPU calculations based on OpenMM [34–36].

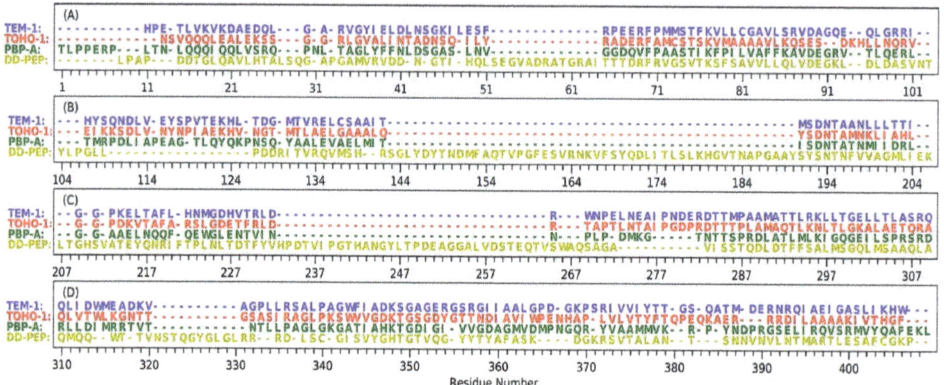

Figure 2. The sequence of four proteins including TEM-1, TOHO-1, PBP-A and DD-transpeptidase with STAMP structural alignment: (**A**) residue indices are from 1 to 103; (**B**) residue indices are from 104 to 206; (**C**) residue indices are from 207 to 309; (**D**) residue indices are from 310 to 409. TEM-1 residues are listed in blue text, TOHO-1 residues are listed in red text, PBP-A residues are listed in green text, DD-transpeptidase residues are listed in yellow text and DD-PEP represents DD-transpeptidase.

2.2. Root-Mean-Square Fluct+uation (RMSF)

RMSF is a parameter to evaluate the fluctuation of conformation for each snapshot of the simulation from the averaged structures:

$$RMSF_i = \left[\frac{1}{T} \sum_{t=1}^{T} \left| r_i(t) - \overline{r_i} \right|^2 \right]^{\frac{1}{2}} \tag{1}$$

2.3. Principal Component Analysis (PCA)

PCA is a widely used dimensionality reduction method for molecular dynamics simulations [37,38]. It could be used to extract the dominant modes of the motion from a trajectory of molecular dynamic simulation. The normal modes for PCA were obtained through diagonalizing the correlation matrix of the atomic position in one trajectory. The correlation matrix element is calculated as:

$$C_{ij} = \frac{c_{ij}}{\sqrt{c_{ii}c_{jj}}} = \frac{\langle r_i r_j \rangle - \langle r_i \rangle \langle r_j \rangle}{\sqrt{\left[(\langle r_i^2 \rangle - \langle r_i \rangle^2)(\langle r_j^2 \rangle - \langle r_j \rangle^2) \right]}} \tag{2}$$

where C_{ij} is the variable of correlation matrix between atoms i and j.

2.4. Configurational Entropy

Entropy is estimated for the simulations using quasi-harmonic approximations based on MD simulations. Quasi-harmonic analysis is calculated by the inversion of the cross-correlation matrix C:

$$F_{ij} = k_B T \left[C^{-1} \right]_{ij} \tag{3}$$

F_{ij} is the element of the force constant matrix F describing the quasi-harmonic potential [39], k_B is the Boltzmann constant and T is the temperature. The configurational entropy S_{config} of the system can be calculated by the vibrational frequency γ_i of the molecule with N atoms:

$$S_{config} = k_B \sum_{i}^{3N-6} \frac{h\gamma_i/k_BT}{e^{h\gamma_i/k_BT} - 1} - \ln\left(1 - e^{-h\gamma_i/k_BT}\right) \tag{4}$$

where h is the Planck constant. The vibrational frequency γ_i in the quasi-harmonic model i of a molecule can be calculated as the solution of the secular equation for angular frequency ω:

$$\det(F - \omega^2 M) = 0 \tag{5}$$

where M is the mass matrix of the molecule [37].

2.5. Time-Structure-Based Independent Component Analysis (t-ICA)

t-ICA [40–43] method was n-dimensional times series, t-ICA is performed by solving generalized eigenvalue problem:

$$\overline{C}F = CKF \tag{6}$$

where $K = diag(k_1,\dots,k_n)$ and $F = (f_1,\dots,f_n)$ are the eigenvalue and eigenvector matrices, respectively. C is the covariance matrix, and $\overline{C}$ is the time-lagged covariance matrix at lagged time τ, which are defined as:

$$C = \langle(x(t) - \langle x(t)\rangle)^t(x(t) - \langle x(t)\rangle)\rangle \tag{7}$$

$$\overline{C} = \langle(x(t) - \langle x(t)\rangle)^t(x(t + \tau) - \langle x(t)\rangle)\rangle \tag{8}$$

Generally, the time-lagged covariance matrix is asymmetric. Avoiding complex numbers in eigenvectors and eigenvalues, a symmetrized time-lagged covariance matrix is given by $\frac{1}{2}(\overline{C} + {}^t\overline{C})$, which is under an assumption of time reversibility of a trajectory. The projected trajectories:

$$a(t) = {}^t(a_1(t), \dots, a_n(t)) = {}^tFx(t) \tag{9}$$

The independent component vectors obtained from t-ICA are uncorrelated and have the maximum autocorrelation value.

2.6. Markov State Models (MSMs)

A MSM [44,45] is used to reduce the complexity of the MD simulations by dividing the phase space into discrete microstates. The discrete microstates are generated by k-means clustering method. Consequently, the transition matrix could be computed, and the element in matrix T_{ij} represents the probability of a microstate starting from microstate i, being transferred to microstate j after the lag time, τ. The dynamics of one system can be decomposed into independent processes represented by the eigenvectors of matrix T. The time scales of the process are computed from the eigenvalues, λ_i, of matrix T as:

$$t_i = -\frac{\tau}{\ln|\lambda_i|} \tag{10}$$

2.7. Hidden Markov Model (HMM)

It was shown that all important mechanistic molecular quantities, both kinetic and thermodynamic, computed by a Markov state model (MSM) are also computable from HMM [25,46,47]. In HMM framework, the basic assumption is that the full phase-space dynamics are Markovian in thermodynamic equilibrium. The dynamics can be projected onto the discrete clusters whose discrete dynamics was observed, which can generate so-called Projected Markov Models (PMMs). If the dynamics are metastable with a number of m slow relaxation processes, and the processes can transfer to the next-faster processes within a separation of timescales, then the PMMs can be approximated by HMM with m hidden states. A maximum likelihood transition matrix could be estimated among hidden states by an adequate lag-time. The lag-time dependent on estimated relaxation timescales is plotted in Supplementary Information (SI), and the lag time used in analysis is selected at the convergence of timescale. Meanwhile, the probability of a microstate belonging to a certain hidden state is estimated.

2.8. Transition-Path Theory

In order to build the transition pathways, two subsets of the state space corresponding to unbound structures (initial states) and bound complex (end states) are defined to investigate the transition processes. All the other states are intermediate states. The committor probability q_i^+ is defined as the probability at state i, in which the system will reach the end state next rather than the initial state. According to the definition, $q_i^+ = 0$ for all i in initial state and $q_i^+ = 1$ for all i in end state. The committor probability for all intermediate states i can be calculated by the following equation:

$$- q_i^+ + \sum_{k \in I} T_{ik} q_k^+ = - \sum_{k \in end\ states} T_{ik} \tag{11}$$

The committor increases from an initial state to an end state. The effective flux $f_{ij} = \pi_i q_i^- T_{ij} q_j^+$, where π_i is the stationary probability when the transition matrix T_τ is normalized. Here, T_τ has a single eigenvector and eigenvalue 1, since it is ergodic (within finite time any state can be reached from any other state), q_i^- is the probability at state i and previously at an initial state. In equilibrium of a molecule, $q^- = 1 - q^+$. For any intermediate states pair i, j, both the f_{ij} and f_{ji} are positive. If the f_{ij}^+ is only considered, $f_{ij}^+ = max\{0,\ f_{ij} - f_{ji}\}$. The total flux can be calculated by:

$$F = \sum_{i \in initial\ state} \sum_{j \notin initial\ state} \pi_i T_{ij} q_j^+ \tag{12}$$

The individual pathways p_i connecting an initial state to an end state can be decomposed from the flux. In equilibrium, a flux can be nonuniquely decomposed into pathways from an initial state to an end state. The decomposition generates a set of pathways p_i along f_i. The f_i provides a relative probability when the set of pathways p_i is considered [24,48]:

$$p_i = \frac{f_i}{\sum_j f_j} \tag{13}$$

In the current study, the PCA of all α-carbon coordinates for four proteins in three states are analyzed by Bayesian HMM [46]. HMM is used to explore the conformational changes of overall structures of four proteins in each state. α-Carbons are backbone carbons, which represent the relative position with reference to other functional groups. Protein α-carbons have been used in many computational methods, such as principal component analysis and recurrence quantification analysis, to elucidate the essential dynamics and functions of proteins [49–51]. In this paper, α-carbon coordinates were subjected to HMM analysis to extract the overall dynamic motions of proteins in three states. To apply HMM, simulations were first grouped into microstates through clustering analysis. Then several macrostates could be identified with appropriate lag time and estimated transition probabilities. The lag-time and number of macrostates for each protein in different states are not unique (time scales dependent on lagged time listed in Supplementary Figures S1–S4). The top two tICs (lag-time 2 ns used for all four proteins) of pairwise distances among heavy atoms of residues in active site (residues are listed in Table 1) were carried out for each protein in different states. MSMs are constructed to elucidate the conformational changes of active site for each protein complex with penicillin G. Perron-cluster cluster analysis (PCCA) was applied to map microstates onto macrostates. The lag-time are selected depending on the implied timescale plots listed in Supplementary Figure S5. The lag time at the end of the x-axis is the adequate lag time (Supplementary Figures S1–S5) used in further construction. PyEMMA version 2.5.6 [25,52–54] was employed to build Markov state Models and Hidden Markov state Model. Other parameters besides lag time in the construction of models are set to their default values.

Table 1. The residues in active site of TEM-1, TOHO-1, PBP-A and DD-transpeptidase.

TEM-1	TOHO-1	PBP-A	DD-Transpeptidase
S70	S70	S61	S62
K73	K73	K64	K65
S130	S130	S122	Y159
N132	N132	N124	N161
N166	A166	L158	A237
K234	K234	K219	H298
S235	T235	T220	T299
A237	S237	D222	T301
G244	N245	G228	T307

3. Results

3.1. Hidden Markov State Models Analysis of Overall Structures

The structures of TEM-1 (PDB ID: 1FQG), TOHO-1 (PDB ID: 1IYQ), PBP-A (PDB ID: 2J8Y) and DD-transpeptidase (PDB ID: 1PWC) were aligned for the comparisons in structure and sequence (Figure 2). All four proteins are subjected to a structural alignment by MultiSeq tool in VMD under STAMP structural alignment algorithm [55,56]. TEM-1 and TOHO-1 both belong to class A β-lactamases and have a very high sequence similarity. PBP-A and DD-transpeptidase belong to the penicillin-binding proteins (PBPs). Although PBPs were reported to share low sequence similarity with β-lactamases [13], PBP-A does show high homology level with TEM-1 and TOHO-1. Comparing to the other three proteins, DD-transpeptidase has several long insertions in its structure. The longest sequence insertion of DD-transpeptidase (residue number 143 to 190 in Figure 1B) was close to the Ω-loop of TEM-1 (residue number 191 to 200 in Figure 1B). The residue numbers listed in the Figure 2 are based on the STAMP structural alignment and do not represent the real residue ordered in the PDB structures.

The acylation mechanism of Class A β-lactamase is divided into two steps, shown in Scheme 1A [57]. In the first step, a proton is abstracted from Ser70 and another proton is transferred to Glu166 via water as a bridge. The nucleophilic (deprotonated Ser70) attacks the carbonyl group of the β-lactam ring. In this step, the tetrahedral intermediate structure is formed. Next, the cleavage of the β-lactam bond is accompanied by Ser130, Lys73 is neural, and Glu166 remains protonated. Then, a proton transfers from Glu166 to Lys73, with the formation of the acyl-enzyme.

The deacylation mechanism of Class A β-lactamase is divided into three steps, shown in Scheme 1B [58]. In the first step, the nucleophilic attack catalyzed by the hydrolytic water molecule occurs on the carbonyl carbon of the acyl-enzyme. And the Glu166 plays as a general base accepting the proton from the water molecule. In the second step, the bond between Ser70-O and the β-latam carbonyl carbon atom is broken. In the third step, the protonation state of the protein is regenerated through the hydrogen transfer from the Glu166 carboxylate to the Ser70-O.

The secondary structures of TEM-1, TOHO-1, PBP-A and DD-transpeptidase are illustrated in Figure 3 with the penicillin G molecule shown as sticks and balls. There are three unique active site residues shared by all four proteins. Through the comparison among the TOHO-1 (1IYQ) [5] and TEM-1 (1FQG) [4] active sites in acyl-intermediate structures, PBP-A (2J8Y) [13] active site binding with penicillin G, and DD-transpeptidase (1PWC) [15] active site binding with penicillin G, three residues are shared among all four proteins. These common active site residues are Ser70, Lys73 and Asn132 in TEM-1, Ser70, Lys73 and Asn132 in TOHO-1, Ser61, Lys64 and Asn124 in PBP-A, and Ser62, Lys65 and Asn161 in DD-transpeptidase. These three common active residues are illustrated in Figure 3.

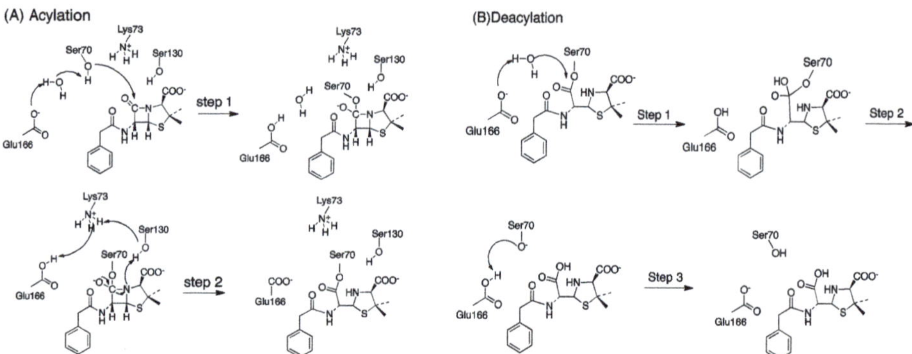

Scheme 1. (**A**) The acylation mechanism of Class A β-lactamases, Step 1: The nucleophilic attack and formation of the tetrahedral intermediate; Step 2: Formation of the acylenzyme from the tetrahedral intermediate (**B**) The deacylation mechanism of Class A β-lactamases, Step 1: The nucleophilic attack by a water molecule and formation of the tetrahedral intermediate; Step 2: The bond-breaking between Ser70-O of the enzyme and the β-latam carbonyl carbon atom; Step 3: Hydrogen atom transfers from Glu166 to Ser70-O and the regeneration of protonation state of the enzyme.

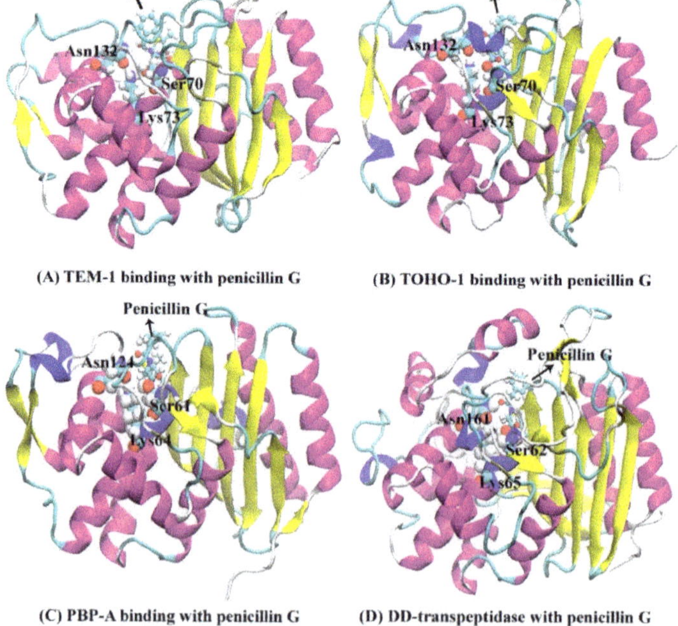

Figure 3. The structures of proteins binding with penicillin G molecule (red sticks): (**A**) TEM-1; (**B**) TOHO-1; (**C**) PBP-A; (**D**) DD-transpeptidase. The binding pockets are identified around three active site residues represented by balls (Ser70, Lys73 and Asn132 for TEM-1; Ser70, Lys73 and Asn132 for TOHO-1; Ser61, Lys64 and Asn124 for PBP-A; Ser62, Lys65 and Asn161 for DD-transpeptidase), penicillin G is illustrated as balls and sticks.

The Hidden Markov state Model (HMM) analyses were carried for TEM-1 in apo, reactant and product states based on the principal component analysis (PCA) of all α carbon coordinates in each state. In each state of TEM-1, the saved snapshots from simulations are projected onto the top two main vectors referred to as the principal component 1 (PC1) and principal component 2 (PC2) space. The projected snapshots are consequently subjected to the *k*-means clustering analysis and divided into 200 microstates on the PC1/PC2 surface. Evaluations of the implied timescale show convergences at lag time 160 ps for the apo state, 200 ps for the reactant state, and 180 ps for the product state (Supplementary Figure S1). The HMM was applied to construct metastable macrostates from the clustered microstates using appropriate lag time for each state.

The HMM analyses of TEM-1 result in three, three, and four metastable macrostates for the apo, reactant and product states, respectively (Figure 4). The metastable macrostates are illustrated in different colors for comparison purpose. In the three states of TEM-1, the macrostates distributed in a similar area on the PCA surface are in the same color, suggesting that these macrostates share a similar dynamic behavior. The transition probabilities among the macrostates for each TEM-1 state were calculated using HMM. For each macrostate, the probability to remain in the current state is overwhelmingly higher than the probability transferring to any other states. Therefore, each macrostate represents a free energy minimum on the surface with kinetic barriers preventing transformations to other macrostates. The simulation of TEM-1 apo state comprises three macrostates (Figure 4A). The macrostate 2 with the largest coverage on the PCA surface is the dominant state among three macrostates. The three macrostates in TEM-1 reactant state (Figure 4B) have an arrangement similar to the apo state, suggesting a similar dynamical behavior for the TEM-1 in these two functional states. Interestingly, the interconversion between macrostates 2 and 3 is extremely unlikely in the reactant state (Figure 4B). This demonstrates the significant impact from the binding with penicillin G on the dynamical properties of TEM-1. The TEM-1 product state simulations were divided into four macrostates (Figure 4C). The distribution of these four macrostates on the PCA surface is significantly different from the apo and reactant states. The interconversion between macrostates 1 and 4 in the product state is extremely unlikely (Figure 4C).

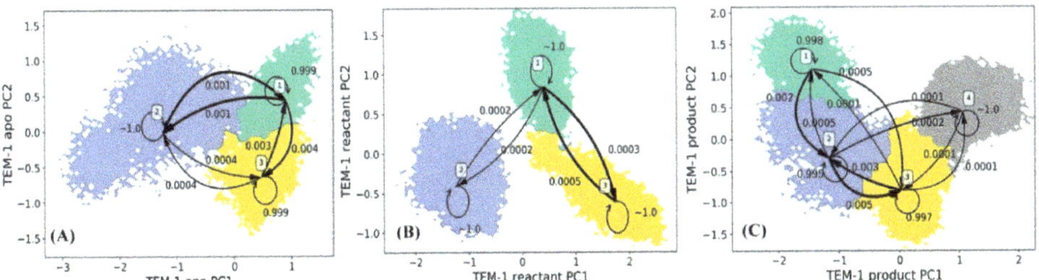

Figure 4. Hidden Markov state model (HMM) of TEM-1 apo (**A**), reactant (**B**), and product (**C**) states. The macrostates resulted from the HMM are based on the 200 microstates generated using *k*-means clustering analysis of TEM-1 simulations in each state. The simulations of each TEM-1 state are projected onto two main vectors referred to as the principal component 1 (PC1) and principal component 2 (PC2) from the principal component analysis (PCA) of all α carbon coordinates in each state simulations. The transition probabilities among macrostates in HMM are also listed. Overall, the probability to remain in the macrostate is overwhelmingly higher than transferring to any other macrostates.

To further understand the dynamical behaviors of TEM-1 in different functional states, the representative structures for each macrostate in apo, reactant and product states of TEM-1 are illustrated in Figure 5. The RMSFs for each macrostate of TEM-1 in three states are plotted in Supplementary Figure S10 to distinguish the key conformational change. Key secondary structures with significantly different RMSF values in three functional states are highlighted in different colors (Figure 5). Penicillin G and its hydrolyzed product as ligands in the reactant and product states are illustrated in space filling mode. Residues 163 to 178 located around the active site in the TEM-1 structure is referred to as Ω loop, which is a critical functional group in the catalytic process of TEM-1 with penicillin G.

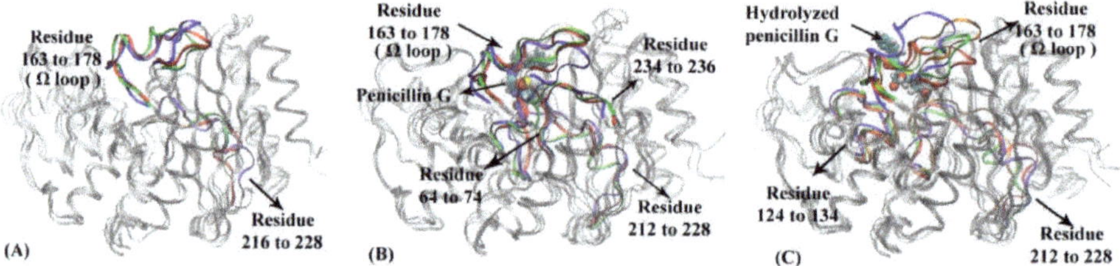

Figure 5. The representative structures of macrostates in three functional states of TEM-1: (**A**) apo state, (**B**) reactant state, (**C**) product state. In TEM-1 apo state (**A**): residue 163 to 178 (Ω loop), residue 216 to 228 are highlighted in macrostate 1 (blue), macrostate 2 (red), macrostate 3 (green). In TEM-1 reactant state (**B**): residues 64 to 74, residues 163 to 178 (Ω loop), residues 212 to 228 and residues 234 to 236 are highlighted in macrostate 1 (blue), macrostate 2 (red), macrostate 3 (green). The penicillin G ligand is illustrated as space filling model in the binding pocket. In TEM-1 product state (**C**): residues 124 to 134, residues 163 to 178 (Ω loop) and residues 212 to 228 are highlighted in macrostate 1 (blue), macrostate 2 (red), macrostate 3 (green), and macrostate 4 (orange). The hydrolyzed penicillin G (product state) is illustrated as space filling model in the binding pocket.

The structural differences among three macrostates in the TEM-1 apo state mainly stem from the Ω loop and residues 216 to 228 (Figure 5A). The residues 216 through 228 also form a loop-like structure, which is located at the distal end of the active site and displays certain flexibility in the apo state.

In the reactant state, the Ω loop and residues 216 to 228, 64 to 74, 234 to 236, and 211 to 215 display significant differences among three macrostates (Figure 5B). Compared to the apo state, the binding with the penicillin G ligand changes the distribution of Ω loop in the reactant state, in favor of conformations closer to the ligand. In addition, the binding with the ligand diminishes the flexibility of loop region residues 212 to 228. Residues 64 to 74 and 234 to 236 as two loops adjacent to the active site, however, show a slightly higher flexibility than in the apo state. Some key catalytic residues for TEM-1, including Ser70, Lys73, Ser235, belong to these regions.

The representative structures of four macrostates in TEM-1 product state are illustrated in Figure 5C. Surprisingly, the Ω loop displays much higher flexibility than both the apo and reactant states. But a loop region of residues 212 to 228 displays conformations similar to the reactant state. Different from the apo and reactant states, another loop region of residues 124 to 134 displays significant conformational changes of TEM-1 in the product state. Interestingly, the residues 130, 131, 132 in this loop is important functional residues in catalytic mechanism [59,60].

The simulations of TOHO-1 in apo, reactant and product states are also projected onto the surfaces formed by top two vectors (PC1 and PC2) from the PCA using all α carbon coordinates for each state. The projected snapshots were consequently subjected to the k-means clustering analysis and divided into 200 microstates based on structural differences. HMM was then applied on the microstates to generate macrostates. The lag time used in the construction of HMM for TOHO-1 is 160 ps for the apo state, 160 ps for the reactant state, 180 ps for the product state as shown in Supplementary Figure S2A–C. The HMM analysis of TOHO-1 resulted into three, four, and five macrostates for the apo, reactant and product states, respectively (Figure 6). The transition probabilities among macrostates in each TOHO-1 functional state is also estimated using HMM and labeled (Figure 6). The probability of each macrostate remaining in itself is much higher than the probabilities of transferring to other macrostates, making each macrostate as a free energy minimum.

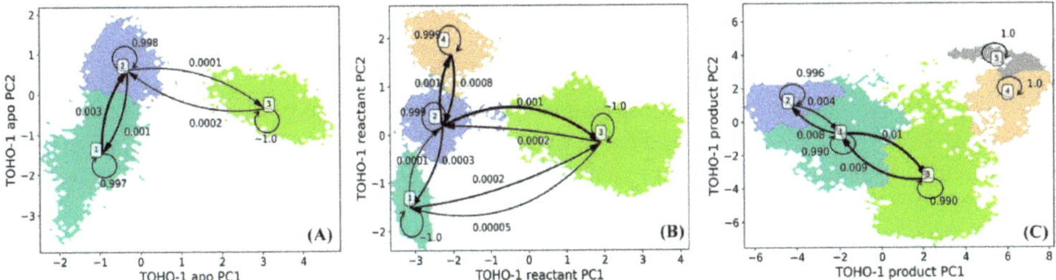

Figure 6. Hidden Markov state model (HMM) of TOHO-1 (**A**) apo, (**B**) reactant, and (**C**) product states. The macrostates resulted from the HMM are based on the 200 microstates generated using *k*-means clustering analysis of TOHO-1 simulations in each state. The simulations of each TOHO-1 state are projected onto two main vectors referred to as the principal component 1 (PC1) and principal component 2 (PC2) from the principal component analysis (PCA) of all α carbon coordinates in each state simulations. The transition probabilities among macrostates in HMM are also listed.

The HMM analysis of TOHO-1 apo state resulted into three macrostates (Figure 6A). The macrostates 1 and 2 are adjacent to each other with relatively high transition probabilities among them. The representative structures of three macrostates are illustrated in Figure 7A and the key conformational changes are highlighted in blue, red and green colors for macrostates 1, 2 and 3, respectively. Surprisingly, the TOHO-1 apo state does not show a significant flexibility. The Ω loop (residues 160 to 178) displays a limited flexibility. The main conformational changes among three macrostates lie in the helix of N-terminus (residues 27 to 45), which swings away from the protein in the macrostate 3 (highlighted in green).

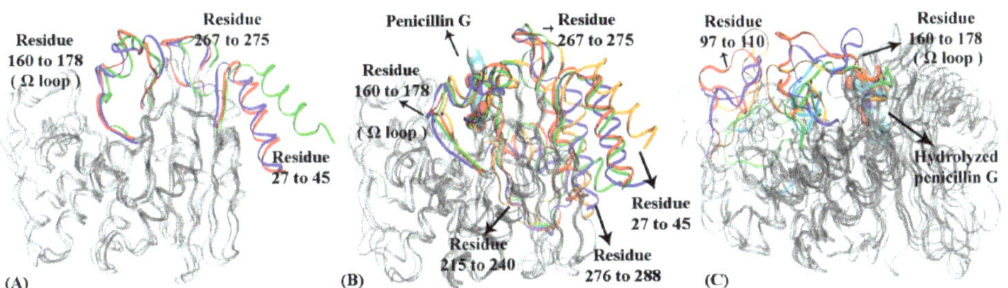

Figure 7. The representative structures of macrostates in three functional states of TOHO-1: (**A**) apo state, (**B**) reactant state, (**C**) product state. In TOHO-1 apo state (**A**) N-terminus helix (residues 27 to 45), a loop of residues 267 to 255 and Ω loop (residues 160 to 178) are highlighted in macrostate 1 (blue), macrostate 2 (red), and macrostate 3 (green). In TOHO-1 reactant state (**B**) N-terminus helix (residues 27 to 45), Ω loop (residues 160 to 178), loop of residues 215 to 240, loop of residues 267 to 275 and loop of residues 275 to 288 are highlighted in macrostate 1 (blue), macrostate 2 (red), macrostate 3 (green) and macrostate 4 (orange); the penicillin G is illustrated in space filling mode at the binding pocket. In TOHO-1 product state (**C**): loop of residues 97 to 110 and Ω loop (residues 160 to 178) are highlighted in macrostate 1 (blue), macrostate 2 (red), macrostate 3 (green), macrostate 4 (orange) and macrostate 5 (cyan). The hydrolyzed penicillin G (product state) is illustrated in the space filling mode.

The distribution of four macrostates in TOHO-1 reactant state (Figure 6B) closely resembles the distribution of the macrostates in TOHO-1 apo state (Figure 6A). Macrostates 1, 2, and 4 in the TOHO-1 reactant state lie in the left-hand side of the PC1-PC2 surface, covering the area corresponding to the macrostates 1 and 2 in the TOHO-1 apo state. The distributions of macrostates 1, 2, and 4 of TOHO-1 reactant state show little overlapping between adjacent states 1 and 2 as well as adjacent states 2 and 4. The representative structures of macrostates 1 to 4 are illustrated in Figure 7B. The secondary structures with significant conformational changes among different macrostates are highlighted in blue (macrostate 1), red (macrostate 2), green (macrostate 3) and orange (macrostate 4). It is interesting that the helix at the N-terminus (residues 27 to 45) in the reactant state shows more flexibility than the apo state. The Ω loop shows the comparable flexibility to the apo state, but there is a loop region (residue 267 to 275) close to the ligand displaying significant conformational changes among different macrostates. In addition, the loop comprising residues 215 to 240 and helix comprising residues 276 to 288 also display higher flexibilities in the reactant state than in the apo state. Overall, the binding with penicillin G seems to increase the overall flexibility and conformational distribution of TOHO-1.

The distribution of five macrostates in TOHO-1 product state (Figure 6C) is dramatically different from the apo and reactant states. The distributions of macrostates 1, 2 and 3 are close to each other and cover the majority of the surface. The transition probabilities between states pairs 1 and 3, 1 and 2 are rather high. Transitions to and from states 4 or 5 are rather rare, and do not lead to meaningful transition probabilities associated with either of these two states, suggesting that these two states are rather isolated in the product state. The representative structures of these five macrostates are illustrated in Figure 7C. Interestingly, the binding mode of the hydrolysis product of penicillin G is quite different from the binding mode in the reactant state, and leads to a different dynamical behavior of the protein. The Ω loop shows a significant flexibility comparing to the apo and reactant states. This is actually similar to the case of TEM-1. Another loop region of residues 97 to 110 of TOHO-1 also shows a higher flexibility than in both apo and reactant states of TOHO-1. Both of these two loops are away from the ligand of product state. On the other hand, the N-terminus (residues 27 to 45) is more rigid in the product state than in the apo and reactant states. In both TEM-1 and TOHO-1 case, the binding with hydrolysis product of penicillin G leads to the dynamical behavior of the protein dramatically different from both the apo and reactant states. These findings indicate the importance of dynamical behavior in different functional states of β-lactamases. It is also worth to point out that the dynamical behaviors of both TEM-1 and TOHO-1 in product state are significantly different from the apo and reactant states. This may suggest that the key differences in dynamical behaviors of β-lactamases in different functional states are important for their catalytic functions.

The HMM analysis was carried out for the simulations of PBP-A in apo, reactant, and product states similar to TEM-1 and TOHO-1, and lead to four macrostates in each functional state of PBP-A (Figure 8). The lag times used for PBP-A three states is are 180 ps for the apo state, 180 ps for the reactant state, 160 ps for the product state as shown in Supplementary Figure S3. In the apo state of PBP-A, macrostates 2, 3, and 4 are adjacent to each other, with significant transition probabilities among them. The macrostate 1 is separated from the macrostates 2, 3, and 4, and is connected to macrostate 2 with a detectable transition probability. The representative structures of four macrostates in PBP-A apo states are illustrated in Figure 9A. Among the macrostates of apo state, the Ω loop (residues 154 to 164) does not show significant conformational changes. But a loop region of residues 96 to 108 at the active site shows significant conformational changes. Despite the overall structural difference, the structural alignment between PBP-A and TEM-1 shows that the Ω loop in PBP-A (residues 152 to 166) aligns well to the Ω loop of TEM-1.

All four macrostates in the reactant state of PBP-A are clustered together with significant transition probabilities among them (Figure 8B). This indicates that the PBP-A bound with penicillin G in reactant state may not be flexible. The representative structures of four macrostates of PBP-A in the reactant state are illustrated in Figure 9B. The binding with the ligand diminishes the flexibility of overall structure of PBP-A, especially the loop region of residues 96 to 108. However, the Ω loop (residues 152 to 166) shows somewhat a higher flexibility in the reactant state than in the apo state. The residues 48 to 64 include a key active site residue Ser 61. Residues 48 to 64 have a relative high flexibility because of a flexible Ser61 and adjacent residues. Although the region does not seem to have high conformational changes in representative structures, the actual RMSFs of this region are higher than other regions as plotted in Supplementary Figure S17.

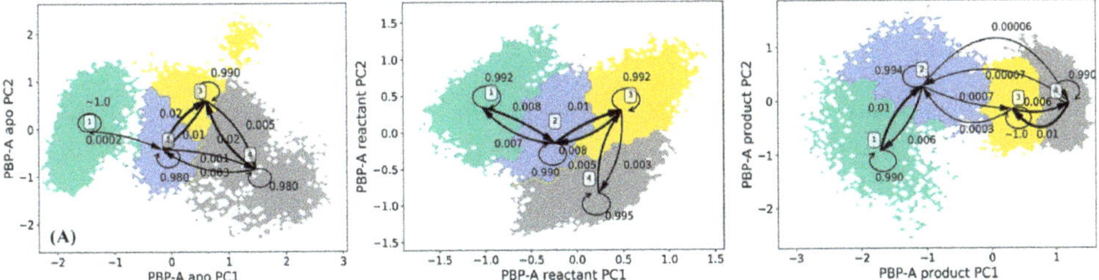

Figure 8. Hidden Markov state model (HMM) of PBP-A (**A**) apo, (**B**) reactant, and (**C**) product states. The macrostates resulted from the HMM are based on the 200 microstates generated using *k*-means clustering analysis of PBP-A simulations in each state. The simulations of each PBP-A state are projected onto two main vectors referred to as the principal component 1 (PC1) and principal component 2 (PC2) from the principal component analysis (PCA) of all α carbon coordinates in each state simulations. The transition probabilities among macrostates in HMM are also listed.

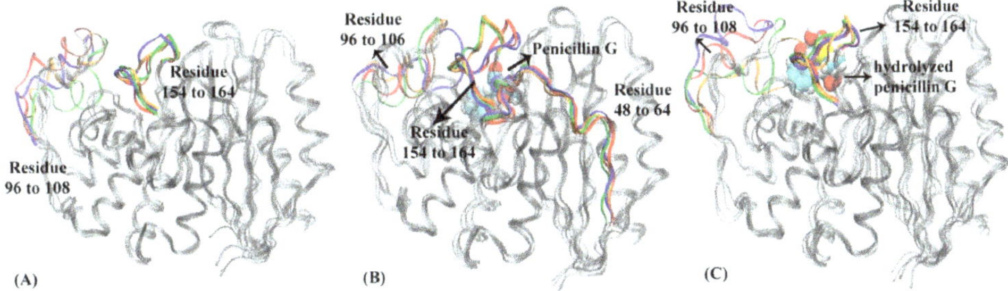

Figure 9. The representative structures of macrostates in three functional states of PBP-A: (**A**) apo state, (**B**) reactant state, (**C**) product state. In PBP-A apo state (**A**): loop of residues 96 to 108 and Ω loop (residues 154 to 164) are highlighted in macrostate 1 (blue), macrostate 2 (red), macrostate 3 (green) and macrostate 4 (orange). In PBP-A reactant state (**B**): residues 48 to 64, loop of residues 96 to 106, and Ω loop of residues 154 to 164 are highlighted in macrostate 1 (blue), macrostate 2 (red), macrostate 3 (green) and macrostate 4 (orange). The penicillin G is illustrated in space filling mode in the binding pocket. In PBP-A product state (**C**): loop of residues 96 to 108 and Ω loop of residues 154 to 164 are highlighted in macrostate 1 (blue), macrostate 2 (red), macrostate 3 (green), macrostate 4 (orange). The hydrolyzed penicillin G (product state) is illustrated in space filling mode in the binding pocket. The hydrolyzed penicillin G (product state) is represented in the space filling mode.

The four macrostates in the product state of PBP-A are also clustered together with significant transition probabilities among them (Figure 8C). This may suggest that the overall structural flexibility of PBP-A in this state is low. The representative structures of four macrostates in the product state are illustrated in Figure 9C. The Ω loop (residues 152 to 166) shows similar conformations to the apo state. Interestingly, the loop of residues 96 to 108 is more flexible than in both apo and reactant states.

The HMM analysis was carried out for the simulations of DD-transpeptidase in the apo, reactant, and product states similar to the other three proteins, and lead to four macrostates in each functional state of PBP-A (Figure 10). The lag times used in the construction of HMM for DD-transpeptidase are 180 ps for the apo state, 160 ps for the reactant state, and 180 ps for the product state as shown in Supplementary Figure S4A–C. In the apo state of DD-transpeptidase, all four macrostates are adjacent to each other with significant transition probabilities among them (Figure 10A). Similar to the PBP-A analysis, this could indicate that the protein does not show a high flexibility in this state. The representative structures of four macrostates in the apo state are illustrated in Figure 11A. Residues 117 to 141 with mixed loop and helix, residues 227 to 243 (corresponding to the Ω loop in TEM-1 under structural alignment in Figure 2C), and the loop of residues 273 to 279 are highlighted with the significant conformational variance in overall DD-transpeptidase structure (Figure 11A). It should be noted that the residues 117 to 141 are located in an insertion of sequence (among residues 110 to 157 as shown in Figure 2B). Residues 273 to 279 are among another insertion of sequence (residues 275 to 284).

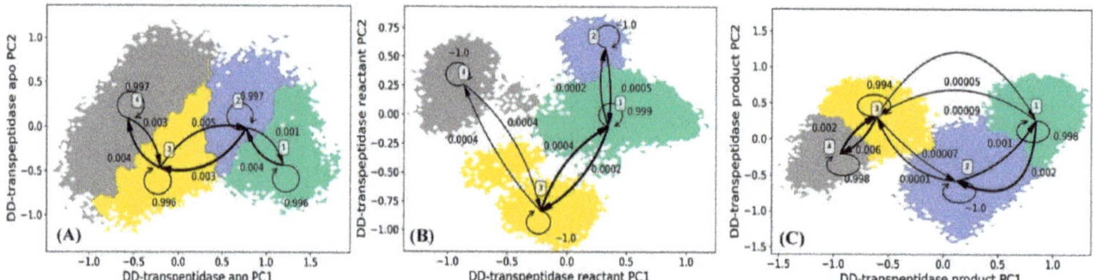

Figure 10. Hidden Markov state model (HMM) of DD-transpeptidase (**A**) apo, (**B**) reactant, and (**C**) product states. The macrostates resulted from the HMM are based on the 200 microstates generated using *k*-means clustering analysis of DD-transpeptidase simulations in each state. The simulations of each PBP-A state are projected onto two main vectors referred to as the principal component 1 (PC1) and principal component 2 (PC2) from the principal component analysis (PCA) of all α carbon coordinates in each state simulations. The transition probabilities among macrostates in HMM are also listed.

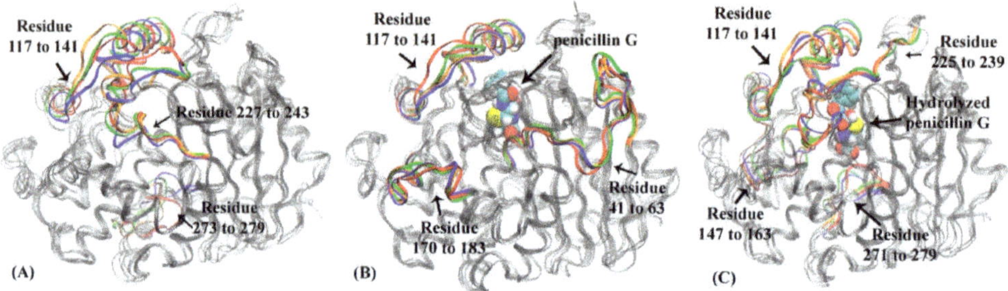

Figure 11. The representative structures of macrostates in three functional states of DD-transpeptidase: (**A**) apo state, (**B**) reactant state, (**C**) product state. In DD-transpeptidase apo state (**A**): residues 117 to 141, residues 227 to 243 and residues 273 to 279 are highlighted in macrostate 1 (blue), macrostate 2 (red), macrostate 3 (green) and macrostate 4 (orange). In DD-transpeptidase reactant state (**B**): residues 41 to 63, residues 117 to 141 and residues 170 to 183 are highlighted in macrostate 1 (blue), macrostate 2 (red), macrostate 3 (green) and macrostate 4 (orange). The penicillin G is represented in space filling mode in the binding pocket. In DD-transpeptidase product state (**C**): residues 117 to 141, residues 147 to 163, residues 225 to 239 and residues 271 to 279 are highlighted in macrostate 1 (blue), macrostate 2 (red), macrostate 3 (green), macrostate 4 (orange). The hydrolyzed penicillin G (product state) is illustrated in the space filling mode.

Comparing to the apo state, the four macrostates in the reactant state of DD-transpeptidase are separated from each other (Figure 10B). The representative structures of these macrostates in the reactant state are illustrated in Figure 11B, and do not show a significant difference from the apo state structures. Two loops (residues 41 to 63 and residues 170 to 183) and residues 117 to 141 (including mixed loop and helix) with significant conformational changes among four macrostates are highlighted. Residues 117 to 141 are in an insertion of sequence in DD-transpeptidase comparing to TEM-1, TOHO-1 and PBP-A structure. Residues 41 to 63 includes a key active site residue Ser62. A loop formed by residues 170 to 183 away from the active site shows a higher flexibility in the reactant state than in the apo state.

Four macrostates in the product state of DD-transpeptidase are rather close to each other on the PCA surface (Figure 10C). The distribution of these four states are similar to both the apo and reactant states, which indicates a similar dynamical behavior in three functional states of DD-transpeptidase. The representative structures of these macrostates are illustrated in Figure 11C. The flexibility displayed by the residues 117 to 141 is similar to those in the apo and reactant states. The other three loops region, residues 147 to 163, residues 225 to 239 and residues 271 to 279 with significant conformational changes do not show high flexibilities either. All the four loops are located around hydrolyzed product of penicillin G as a ligand.

3.2. Analysis of Active Site Structures Using Markov State Models

The above analyses strongly indicate that TEM-1 and TOHO-1 as two Class A β-lactamases share a similar dynamical behavior in different catalytic functional states, including binding with either reactant or product and the apo state as well. PBP-A and DD-transpeptidase, on the other hand, share a similar dynamical behavior in different catalytic functional states. All four proteins share a similar catalytic cavity. The active site residues of each protein are listed in Table 1. To further analyze and directly compare the dynamical behaviors of these proteins related to their catalytic activities, the active site of each protein combining with the reactant penicillin G are subjected to analysis using MSM based on t-ICA. The lag times of the MSM for four proteins are 3 ns as shown in Supplementary Figure S5.

The main focus is the dynamical process of binding between the protein active site and the ligand. During the simulations of all four proteins, the escaping of the ligand from the active site was observed. Only the simulations of reactant state for each protein were subjected to the analysis. Therefore, the binding process of the ligand from the active site in each protein could be characterized as macrostates generated from the MSM. The distribution of macrostates in the reactant state simulations of each protein are plotted in Figure 12.

Such an analysis reveals dominant transition pathways of the system of interest. For example, states sequence [2–5] in Figure 12A represents a transition pathway 4→5→2→3 with probability 0.871, with state 4 is an initial state and state 3 is an end state. The representative structures of active site combining with reactant for macrostates generated from the MSM of trajectories are illustrated in Figure 13. These representative structures are divided into the initial state (unbound state), intermediate states, end state (bound state) and trapped states. The trapped states are those states that are terminal states without being on the transition pathway connecting the bound and unbound states. It should be noted that the unbound states are those structures that are the most different from the bound structure, and are not expected to be a completely dissociated state between the active site and ligand. The trapped states can be described as local minima on kinetic energy surface, separated by energy barrier from the main basin. Such trapped states are rarely visited, and are stable for significant amount of time if they are actually visited [61].

Based on the transition probabilities among these macrostates, one could identify the most probable transition pathways connecting the bound and unbound states. For example, the most probable transition pathway from the unbound state to intermediate state to the bound state for TEM-1 binding with penicillin G is macrostates 4→5→2→3 with transition probability as 0.871 (Figure 12A). The probability is calculated using the committor probability described in transition-pathway theory. The representative structures for macrostates 2–5 illustrated in Figure 13A demonstrate the conformational change of active site with the ligand throughout the binding process. There are other possible transition pathways, including 4→2→6→3 and 4→2→3, which bear much lower transition probabilities. In the macrostate 4 (initial state), the penicillin G resides at the cavity of TEM-1 with a weak interaction with Ala237. In the intermediate states, residue Asn132 in the macrostates 2, 5 and 6 moves opposite to the penicillin G and creates additional space for the ligand. This behavior of Asn132 was reported in an experimental paper [59]. In the six macrostates of TEM-1, residues Ala237 and Lys73 have significant conformational changes. Ala237 moves closer to the ligand. From the initial state to the end state, residue Lys73 gradually moves to the back side and closer to Ser70.

The most probable transition pathway of TOHO-1 binding with penicillin G is identified through MSM as macrostates 5→4→2→1 (Figure 12B). This transition pathway is illustrated in Figure 13. The trapped state 3 seems to be similar to the bound state. Overall, the active site of TOHO-1 reactant state does not show a high flexibility. Only Asn132 has a significant conformational change in this case. Similar to TEM-1, the binding pocket of TOHO-1 is broadened in the intermediate states (macrostates 2 and 4) by the rotation of sidechain of residue Asn132.

The macrostates distributions of active site with penicillin G in PBP-A cover a wide range of t-ICA surface constructed by the two main vectors. The significant transition probabilities among multiple macrostates lead to multiple transition pathways between the bound and unbound state (Figure 11C). The most probable transition pathway is $1 \to 5 \to 6$. The second most probable transition pathway is $1 \to 2 \to 6$. The representative structures are illustrated in Figure 13. Other transition pathways are less probable. It is interesting that the transition pathway $1 \to 5 \to 2 \to 6$ is less probable than the pathway as $1 \to 5 \to 2 \to 4 \to 6$, because of the higher transition probability for transition of $2 \to 4 \to 6$ than the one for transition of $2 \to 6$. The main conformational changes along the transition pathway are from Asn124 and Asp222, which correspond to Asn132 and Ala237 in TEM-1 active site. Residue Asn124 has a similar motion to provide space for the ligand in the intermediate states. However, residue Asp222 provides a closer interaction with penicillin G than Ala237 in TEM-1.

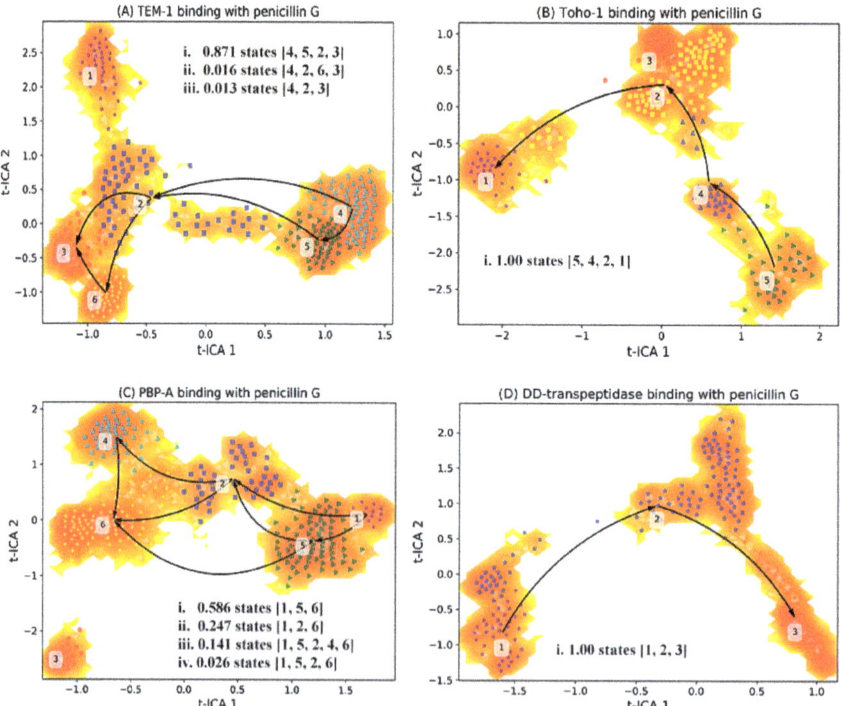

Figure 12. Markov state model analysis of reactant state simulations for (**A**) TEM-1, (**B**) TOHO-1, (**C**) PBP-A, (**D**) DD-transpeptidase. The simulations are subjected to t-ICA to identify the top two vectors (t-ICA 1 and t-ICA 2), upon which the simulations are projected. Based on the representative structures, different macrostates are identified as initial state (unbound state), intermediate states, end state (bound state) and trapped states. It should be noted that the trapped states are terminal states without being on the transition pathway connecting the bound and unbound states. The unbound states are those structures that are the most different from the bound structure, and are not expected to be a completely dissociated state between the active site and ligand.

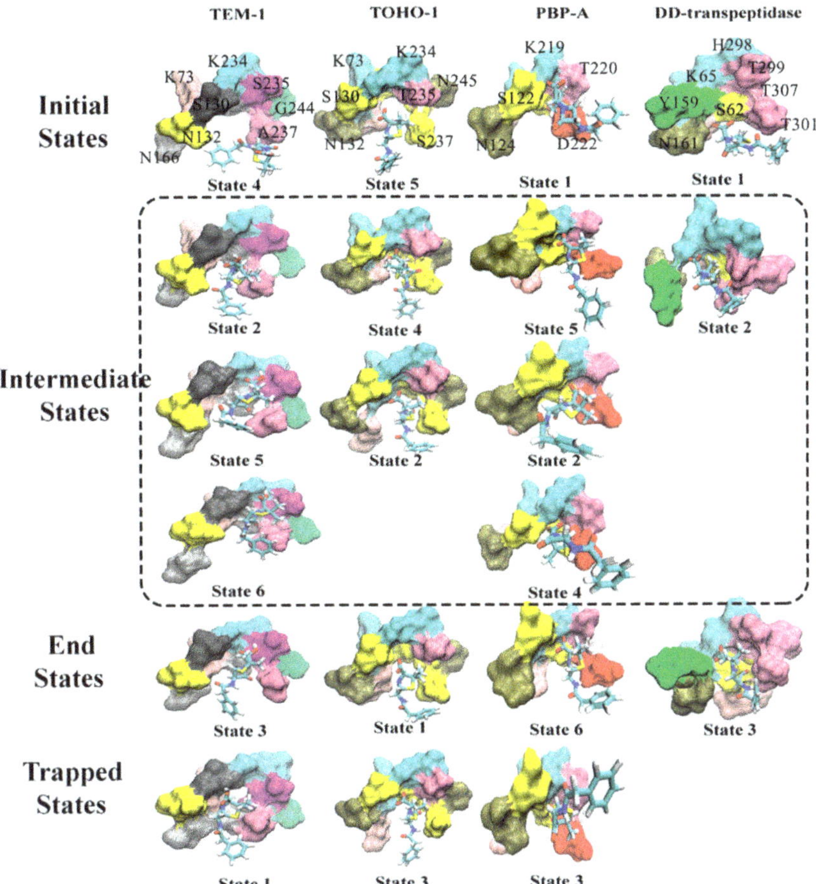

Figure 13. The representative structures of active site of TEM-1, TOHO-1, PBP-A and DD-transpeptidase with penicillin G as a ligand corresponding to the macrostates in Figure 12. The macrostates are divided into four states: initial state (unbound state), intermediate states, end state (bound state) and trapped states. The trapped states are those states that are terminal states without being on the transition pathway connecting the bound and unbound states. The unbound states are those structures that are the most different from the bound structure, and are not expected to be a completely dissociated state between the active site and ligand. The residues in each active site are illustrated as surface in different colors, and penicillin G molecule is illustrated in stick model. The residues in TEM-1 active site include S70, K73, S130, N132, N166, K234, S235, A237, G244; the residues in TOHO-1 active site include S70, K73, S130, N132, A166, K234, T235, S237, N245; the residues in PBP-A active site include S61, S64, S122, N124, L158, K219, T220, D222, G228; the residues in DD-transpeptidase active site include S62, K65, Y159, N161, E237, H298, T299, T301, T307. The residue names are labeled in the initial states, some residues hidden in the back are not labeled.

There are only three macrostates generated for the active site of DD-transpeptidase with penicillin G (Figure 12D). The most probable transition pathway is going through the sole intermediate state. The similarity among the three macrostates of DD-transpeptidase active site (Figure 13) agrees with the observation of the low flexibility of DD-transpeptidase in the reactant state. This is similar to the active site of TOHO-1. Upon the ligand binding, an active site residue Tyr159 has a rotational motion from the initial state to the intermediate state. In the end state, Tyr159 returns to the conformation as in the initial state.

Both the active sites of TEM-1 and PBP-A show a significant flexibility with multiple transition pathways connecting the bound and unbound states. The TEM-1 and PBP-A have similar t-ICA distributions, and TOHO-1 and DD-transpeptidase simulations have similar narrow arched t-ICA surfaces, but the overall distributions of TEM-1 and TOHO-1 are similar to each other, and PBP-A shares similar layouts with DD-transpeptidase.

3.3. Atomic Distance

The analyses of whole protein simulations show that the TEM-1 and TOHO-1 as Class A β-lactamase share similar overall dynamical behaviors, and PBP-A and DD-transpeptidase as PBPs share similar overall dynamical behaviors. But the above analyses of active site dynamics reveal that TEM-1 and PBP-A active sites are more flexible than the active sites of TOHO-1 and DD-transpeptidase. As pointed out earlier in this study, TEM-1, TOHO-1, PBP-A and DD-transpeptidase share three unique residues, including serine, lysine and asparagine, at their active site. [4,5,13,15] To further compare the dynamical behaviors related to their catalysis, the averaged distances among heavy atoms of three residues (in total 23 heavy atoms) and penicillin G (23 heavy atoms) are illustrated as a heat map in Figure 14. The structures of penicillin G and three conserved residues in active site are shown in Figure 15. All the atom numbers on penicillin G, serine, lysine and asparagine structures are listed in Supplementary Scheme S1.

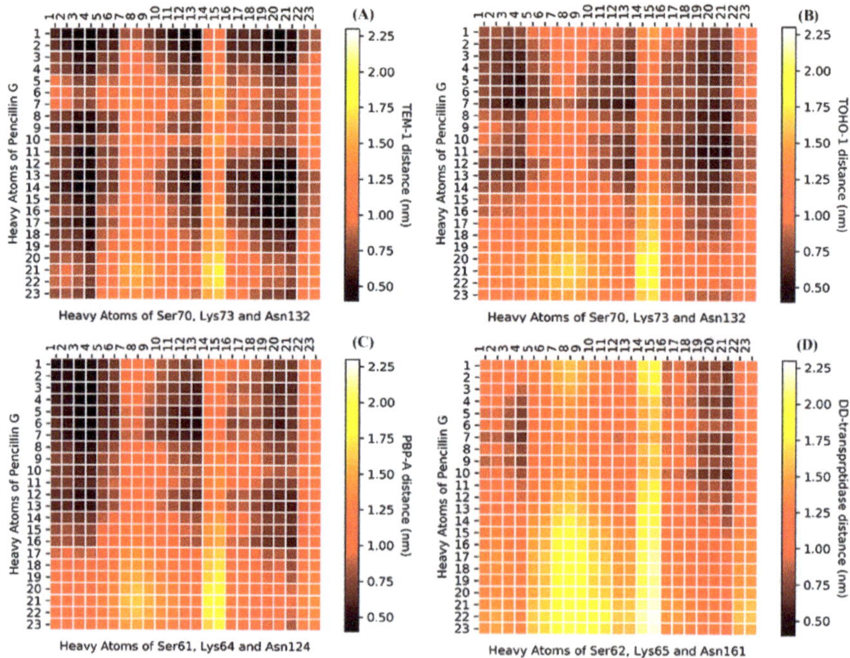

Figure 14. The heat-map of averaged distance (nm) among the heavy atoms of penicillin G and three conserved residues at the active site (**A**) TEM-1: Ser70, Lys73 and Asn132; (**B**) TOHO-1: Ser70, Lys73 and Asn132; (**C**) PBP-A: Ser61, Lys64 and Asn124; (**D**) DD-transpeptidase: Ser62, Lys65 and Asn161. The color bars are listed on the right of the heat map from dark red to light yellow. (The atom numbers of both heavy atoms of penicillin G and three residues are listed in Supplementary Scheme S1, Tables S1 and S2).

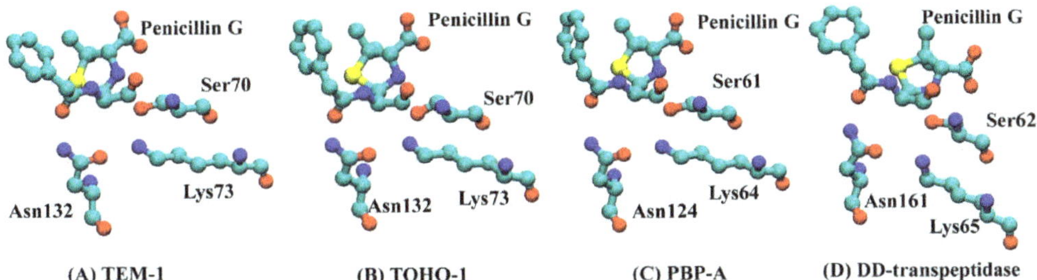

Figure 15. The structures of penicillin G and three conserved residues at active site (Serine, Lysine and Asparagine) in (**A**) TEM-1: Ser70, Lys73 and Asn132; (**B**) TOHO-1: Ser70, Lys73 and Asn132; (**C**) PBP-A: Ser61, Lys64 and Asn124; (**D**) DD-transpeptidase: Ser62, Lys65 and Asn161. The hydrogen atoms are not plotted for clarity.

The darker color means a short distance, and lighter color represents a long distance. The distance heat map could be viewed as a 2D fingerprint for dynamics of this three residues triad in each protein. The bird eye views of TEM-1, TOHO-1 and PBP-A are similar. Although all four heatmaps display overall similar pattern, the heatmaps of TEM-1 and PBP-A are more similar to each other, while the heatmaps of TOHO-1 and DD-transpeptidase are more similar to each other.

3.4. Configurational Entropy

The configurational entropies of four proteins based on the simulations in different states are calculated and plotted in Figure 16.

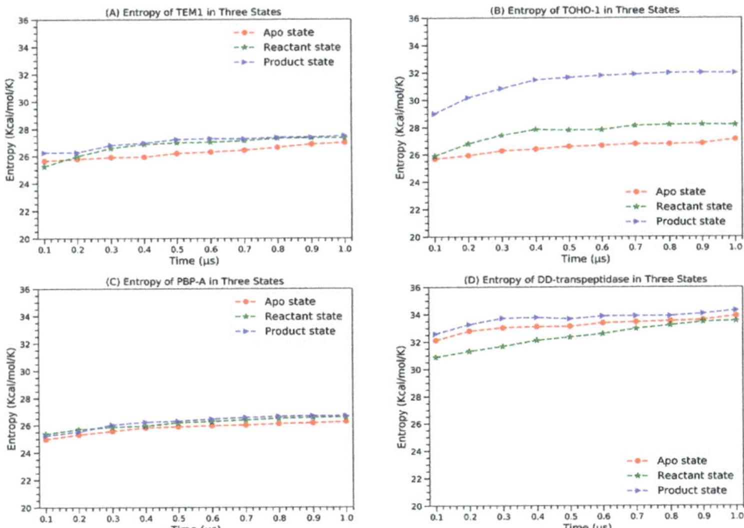

Figure 16. The configurational entropies of four proteins in apo state (red circles), reactant state (green stars) and product state (blue triangles), (**A**) TEM-1, (**B**) TOHO-1, (**C**) PBP-A and (**D**) DD-transpeptidase.

All the proteins display clear convergence tendency in each state. TEM-1 and PBP-A are similar to each other in a way that the apo state has the lowest entropy and the reactant and product states have very close entropies. TOHO-1 and DD-transpeptidase are significantly different. The apo state of TOHO-1 also has the lowest entropy. The entropy of TOHO-1 reactant is marginally higher than the apo state. But the entropy of TOHO-1 product state is significantly higher than both apo and reactant state. Therefore, the entropy could be a main driving factor for the TOHO-1 catalysis against various ligands. Different from other three proteins, DD-transpeptidase reactant state has the lowest entropy. Both reactant and product states entropies are only slightly higher with the product state entropy as the highest. Overall, it is unlikely that entropy plays dominant role in the functions of TEM-1, PBP-A and DD-transpeptidase.

4. Discussion

In this study, TEM-1, TOHO-1, PBP-A and DD-transpeptidase in the apo, reactant, product states were subjected to MD simulations and detailed analyses. Their dynamical behaviors are impacted differently by the presence of penicillin G as a ligand in the reactant state and the hydrolyzed penicillin G as a ligand in the product state. Although the catalytic and dynamic mechanisms of β-lactamases and PBPs in reactant state were investigated intensively, the importance of dynamic behaviors of β-lactamases and PBPs in product state is still underestimated. It has been proposed that DD-transpeptidase could be closely related to a common ancestor of modern PBPs and β-lactamases [23]. The current study provides dynamical information related to the evolutionary relations between β-lactamases and PBPs represented by the target four proteins.

A common Ω loop at the active site for class A β-lactamases has been recognized as a crucial structure for their catalysis against antibiotics [5,62]. One study of TEM-1 using MD simulations and NMR showed that TEM-1 is a rigid structure. The main function of Ω loop is to broaden the catalytic cavity through its fluctuation [63]. Our study provides additional insight into the dynamical behavior of TEM-1 Ω loop as it displays more flexibility in the product state than in the apo and reactant states (Figure 5). Meanwhile, the residues 124 to 134 region also shows significant conformational changes only in TEM-1 product state, and is more stable in reactant and apo state. Therefore, this region could be as a potential mutagenesis position to increase the flexibility of Ω loop. This unique feature of Ω loop is also shared with TOHO-1 (Figure 6). However, both PBP-A and DD-transpeptidase do not show this particular feature of Ω loop. For both proteins, the Ω loop or the structure corresponding to the Ω loop do not show high flexibility in the product states (Figures 9 and 11). Therefore, it could be speculated that the high Ω loop flexibility in the product state could be one of key properties of β-lactamases developed during the evolution from ancient PBPs. A more general observation of these four proteins is that both the apo and reactant states of TEM-1 and TOHO-1 have similar flexibility which is lower than their product states. But all three states of PBP-A and DD-transpeptidase have similar flexibility. Considering the catalytic function of β-lactamases against antibiotics, it could be suggested that higher flexibility in the product state is auxiliary to maintain an appropriate turnover rate.

Residues 218 to 224 form a helix in TEM-1 and is referred to as helix 11. This helix 11 is proposed as an allosteric site in TEM-1 related to its thermal stability [64]. A wider range of residues 216 to 228 enclosing helix 11 display significant flexibility in all three states of TEM-1. In both reactant and product states, this flexible region extends to residue 212 towards the catalytic active site of TEM-1. This observation also supports the allosteric behavior of the helix 11. The Leu220 in this region was proposed to interact with substrates to improve binding. The Val216 also has an auxiliary role to anchor a structurally conserved water molecule [65,66]. Therefore, the region of residues 216 to 228 plays a critical role with substrate binding. In the reactant and product state, this extended flexible region (include up to residue 212) reveals that the dynamics of residues 212 to 215 are impacted by substrate binding interactions. Therefore, residues 212 to 215 could be potential mutagenesis targets to alter the catalytic activities of this protein. Ser70 is a critical residue for TEM-1 catalytic activity, and is located on the loop of residues 64 to 74. The Ser70 is covalently bound to the ring opening intermediate of Penicillin G as a product of the acylation step [4,26]. Interestingly, the loop of residues 64 to 74 display significant flexibility only in the reactant state of TEM-1 (Figure 5), showing its unique role for catalysis in the reactant state.

In TOHO-1, residues 234, 235, and 236 are recognized as the catalytically important KTG sequence [5,67]. Residues 215 to 240 display significant flexibility only in the reactant state of TOHO-1

(Figure 7), also showing its unique role for catalysis in the reactant state. A Ser237Ala mutation in this region can significantly change the catalytic efficiency of this protein. Besides, the residue 240 could affect catalytic efficiencies of many substrates [67]. Furthermore, we propose that other residues in the region of residues 215 to 240 could also serve as mutagenesis targets to adjust the catalytic efficiencies of TOHO-1 against various antibiotics. As an extended-spectrum β-lactamase, TOHO-1 does show higher catalytic activities against more antibiotics than TEM-1. Our analysis shows that TOHO-1 in its reactant state display the higher flexibility than both the apo and product states, which could be correlated with the wide range of ligands profile associated with TOHO-1.

Comparing to other extended-spectrum β-lactamases, TOHO-1 contains a unique residue Arg274 on residues 266 to 288 helix. When binding with penicillin G as a substrate, Arg274 was forced out of the active site since it obstructs the binding pocket of TOHO-1 [5]. This could be one of the reasons that residues 266 to 288 helix display significant flexibility in the reactant state of TOHO-1. Also in TOHO-1, residue Tyr105 was reported to be involved in active site binding with substrate [68]. It was also reported in the same study that Tyr105 displays a single conformation in apo state and two conformations in complexed state. Tyr105 is located on the loop region of residues 97 to 110, which displays significant flexibility only in the product state (Figure 6C).

However, not all the observations about TEM-1 and TOHO-1 in this study agree with the experimental study well. An NMR study of TEM-1 in its apo state identifies residues 124 to 134 with important chemical shift difference [69]. This region, however, displays more prominent flexibility in the product state than in the reactant and apo states (Figure 5). The increased flexibility is reflected in one of the macrostates of TEM-1 in product state (Supplementary Figure S12), and is induced by the binding with the penicillin G hydrolysis product. In addition, the overall RMSFs on total 1 μs simulations of TEM-1 in apo, reactant and product states show similar flexibilities (Supplementary Figure S6). Residues 124 to 134 indeed have fluctuations in TEM-1 apo state, which corresponds to the chemical shift difference in NMR study. However, the flexibility is more observable in the macrostate 2 of product state.

In PBP-A, two loops forming the active site, residues 96 to 108 and Ω loop (residues 154 to 164) display significant flexibilities in all functional states (Figure 9). It was reported that the loop of residues 96 to 108 in PBP-A displays different conformations from the corresponding loop of residues 100 to 115 in TEM-1 [13]. Specifically, Glu104 and Tyr105 in TEM-1 are not well aligned with the equivalent residues Glu96 and Ala97 with an amino acid insertion in PBP-A. This structural difference could be the main reason for the difference of dynamical behavior between PBP-A and TEM-1, which may contribute to the significant differences between the catalytic activity and profiles of these two enzymes.

DD-transpeptidase as the other PBP in this study displays a similar dynamical behavior to PBP-A with similar flexibilities of key secondary structures around the active site. Both Ser62 and Lys65 in DD-transpeptidase are catalytically important residues [15]. The loop of residues 41 to 63 containing residue Ser62 shows moderate flexibility only in the reactant state. In addition, Tyr159 was also proposed as a key residue in acylation step [70,71]. The loop of residues 147 to 163 shows a moderate flexibility only in the product state.

TEM-1 and TOHO-1 in reactant state have increased flexibilities of the residues in the active site than in the apo state, because of the interactions between penicillin G and residues in active site. Starting with the binding with the cavity of protein, the penicillin G ligand gradually leaves the active site after long time simulations, resulting in the observed conformations in initial states, intermediate states, end states and trapped states (Figure 13). Unlike β-lactamases, both PBP-A and DD-transpeptidase do not catalyze wide range of hydrolysis reactions against various antibiotics. Therefore, these enzymes do not display significantly different flexibility among apo, reactant, and product states. In addition, due to the larger size of DD-transpeptidase than other three proteins, the binding with a ligand may lead to smaller impact on protein flexibility.

Further analysis of active site structures combining with the penicillin G as a ligand leads to an unexpected similarity between TEM-1 and PBP-A as well as between TOHO-1 and DD-transpeptidase.

We propose that the dynamic properties of the catalytic cavity of TEM-1 are closer to PBP-A than TOHO-1 in terms of their evolution. And the catalysis related dynamics of TOHO-1 has a closer relationship with DD-transpeptidase than with TEM-1. Additional analysis of three residues serine, lysine and asparagine at active site shared by all four proteins leads to some similarities among TEM-1, TOHO-1, and PBP-A, other than DD-transpeptidase. Given that the crystal structures of TEM-1, TOHO-1, and PBP-A are aligned well to each other, while DD-transpeptidase has numerous sequence insertions comparing to the other three proteins (Figure 2), these comparisons do shed light onto the relations among dynamical behaviors of these proteins in different functional states and the functional and evolutionary relations among class A β-lactamases and PBPs.

5. Conclusions

In this study, the hidden Markov state model was used to analyze the molecular dynamics simulations of two class A β-lactamases, TEM-1 and TOHO-1, and two penicillin binding proteins, PBP-A and DD-transpeptidase. Both principal component analysis and time-lagged independent component analysis were employed as dimensionality reduction methods for the projection of simulations onto reduced dimensions of space. The analysis of dynamical behaviors of overall structures of four proteins agrees with the overall structural comparison in that the key loop structures around the active site in TEM-1 and TOHO-1 displays lower flexibility in the apo and reactant states than in the product state. Likely due to its wider spectrum of ligands, TOHO-1 displays higher overall flexibility in its reactant state than in the apo and product states. Both PBP-A and DD-transpeptidase display consistent flexibility in all three states, agreeing with their specific catalytic functions. Additional analysis of dynamical behavior of active sites complexed with penicillin G in these four proteins reveals that the active sites are rather flexible in TEM-1 and PBP-A with multiple transition pathways between bound and unbound states. Further analysis of three key catalytic residues shared among four proteins indicates similarity of dynamical property among TEM-1, TOHO-1, and PBP-A. All the dynamics analyses presented in this study show the complication of evolutionary relations between β-lactamases and PBPs, and support the notion that protein dynamics may play a significant role in and are characterized by the catalytic functions of these enzymes.

Supplementary Materials: The following are available online at http://www.mdpi.com/1099-4300/21/11/1130/s1, Figure S1: Estimated relaxation timescale based on different lag time for TEM-1. (A) Apostate simulations; (B) Reactant state; (C) Product state. The relaxation timescales are estimated based on transition probabilities among different microstates regarding with the different lag time as interval for analysis, Figure S2: Estimated relaxation timescale based on different lag time for TOHO-1. (A) Apo state simulations; (B) Reactant state; (C) Product state. The relaxation timescales are estimated based on transition probabilities among different microstates regarding with the different lag time as interval for analysis, Figure S3: Estimated relaxation timescale based on different lag time for PBP-A. (A) Apo state simulations; (B) Reactant state; (C) Product state. The relaxation timescales are estimated based on transition probabilities among different microstates regarding with the different lag time as interval for analysis, Figure S4: Estimated relaxation timescale based on different lag time for DD-transpeptidase. (A) Apo state simulations; (B) Reactant state; (C) Product state. The relaxation timescales are estimated based on transition probabilities among different microstates regarding with the different lag time as interval for analysis, Figure S5: Estimated relaxation timescales based on different lag time for active site binding with penicillin G in reactant states of: (A) TEM-1; (B) TOHO-1; (C) PBP-A; (D) DDtranspeptidase. The relaxation timescales are estimated based on transition probabilities among different microstates regarding with the different lag time as interval for analysis, Figure S6: The RMSFs for 1 μs TEM-1 simulations in apo, reactant and product states, Figure S7: The RMSFs for 1 μs TOHO-1 simulations in apo, reactant and product states, Figure S8: The RMSFs for 1 μs PBP-A simulations in apo, reactant and product states, Figure S9: The RMSFs for 1 μs DD-transpeptidase simulations in apo, reactant and product states, Figure S10: The RMSFs for three macrostates in TEM-1 apo state simulations. Region A1 represents residues 163 to 178 (Ω loop), Region A2 represents residues 216 to 228, Figure S11: The RMSFs for three macrostates in TEM-1 reactant state simulation. Region A1 represents residues 64 to 74, region A2 represents residues 163 to 178 (Ω loop), region A3 represents residues 212 to 228, and region A4 represents residues 234 to 236, Figure S12: The RMSFs for four macrostates in TEM-1 product state simulations. Region A1 represents residues 124 to 134, region A2 represents residues 163 to 178 (Ω loop), and region A3 represents residues 212 to 228, Figure S13: The RMSFs for three macrostates of TOHO-1 apo state simulations. Region A1 represents residues 27 to 45, and region A2 represents residues 160 to 178 (Ω loop), Figure S14: The RMSFs for four macrostates in TOHO-1 reactant state simulations. Region A1 represents residues 27 to 45, region A2 represents residues 160 to 178 (Ω loop), region A3 represents residues 215 to 240, and region A4 represents residues 266 to 288, Figure S15: The RMSFs for five macrostates in TOHO-1 product state simulations. Region A1 represents residues 97 to 110, and region A2 represents residues 160 to 178 (Ω loop), Figure S16: The RMSFs for four macrostates in PBP-A apo state simulation. Region A1 represents residues 96 to 108, and region A2 represents residues 154 to 164, Figure S17: The RMSFs for four macrostates in PBP-A reactant state simulations. Region A1 represents residues 48 to 64, region A2 represents residues 96 to 106, and region A3 represents residues 154 to 164 (Ω loop), Figure S18: The RMSFs for four macrostates in PBP-A product state simulations. Region A1 represents residues 96 to 108, and region A2 represents residues 154 to 164 (Ω loop), Figure S19: The RMSFs for four macrstates in DD-transpeptidase apo state simulations. Region A1 represents residues 117 to 141, region A2 represents residues 227 to 243, and region A3 represents residues 273 to 279, Figure S20: The RMSFs for four macrostates in DD-transpeptidase reactant state simulations. Region A1 represents residues 41 to 63, region A2 represents residues 117 to 141, and region A3 represents residues 170 to 183, Figure S21: The RMSFs for four macrostates in DD-transpeptidase product state simulations. Region A1 represents residues 117 to 141, region A2 represents residues 147 to 163, region A3 represents residues 225 to 239, and region A4 represents residues 271 to 279, Scheme S1: Penicillin G and three residues at active site (Serine, Lysine and Asparagine) shared by TEM-1, TOHO-1, PBP-A and DD-transpeptidase. The atomic symbol with sequence numbers are corresponding to the symbol used atomic distances in heat map, Table S1: The atomic numbers used in averaged distance heatmap for the heavy atoms of Serine, Lysine and Asparagine (1-23) shared among all four proteins, Table S2: The atomic numbers used in averaged distance heatmap, the heavy atoms in penicillin G (1-23).

Author Contributions: Formal analysis, F.W.; investigation, F.W. and H.Z.; resources, F.W. and H.Z.; data curation, F.W.; writing—Original draft preparation, F.W.; writing—Review and editing, P.T., X.W. and F.W.; visualization, F.W.; supervision, P.T.; project administration, P.T.; funding acquisition, P.T.

Funding: This material is based upon work supported by the National Science Foundation under Grant No. (1753167).

Acknowledgments: Computational time was provided by Southern Methodist University's Center for Scientific Computation.

Conflicts of Interest: The authors declare no conflict of interest.

References

1. Hall, B.G.; Barlow, M. Evolution of the serine β-lactamases: Past, present and future. *Drug Resist. Updates* **2004**, *7*, 111–123. [CrossRef] [PubMed]
2. Coates, A.; Hu, Y.; Bax, R.; Page, C. The future challenges facing the development of new antimicrobial drugs. *Nat. Rev. Drug Discov.* **2002**, *1*, 895–910. [CrossRef] [PubMed]
3. Medeiros, A.A. Evolution and Dissemination of β-Lactamases Accelerated by Generations of β-Lactam Antibiotics. *Clin. Infect. Dis.* **1997**, *24*, S19–S45. [CrossRef] [PubMed]
4. Strynadka, N.C.J.; Adachi, H.; Jensen, S.E.; Johns, K.; Sielecki, A.; Betzel, C.; Sutoh, K.; James, M.N.G. Molecular structure of the acyl-enzyme intermediate in β-lactam hydrolysis at 1.7 Å resolution. *Nature* **1992**, *359*, 700–705. [CrossRef]
5. Shimamura, T.; Ibuka, A.; Fushinobu, S.; Wakagi, T.; Ishiguro, M.; Ishii, Y.; Matsuzawa, H. Acyl-intermediate Structures of the Extended-spectrum Class A β-Lactamase, Toho-1, in Complex with Cefotaxime, Cephalothin, and Benzylpenicillin. *J. Biol. Chem.* **2002**, *277*, 46601–46608. [CrossRef]
6. Tomanicek, S.J.; Blakeley, M.P.; Cooper, J.; Chen, Y.; Afonine, P.V.; Coates, L. Neutron Diffraction Studies of a Class A β-Lactamase Toho-1 E166A/R274N/R276N Triple Mutant. *J. Mol. Biol.* **2010**, *396*, 1070–1080. [CrossRef]
7. Tomanicek, S.J.; Standaert, R.F.; Weiss, K.L.; Ostermann, A.; Schrader, T.E.; Ng, J.D.; Coates, L. Neutron and X-ray Crystal Structures of a Perdeuterated Enzyme Inhibitor Complex Reveal the Catalytic Proton Network of the Toho-1 β-Lactamase for the Acylation Reaction. *J. Biol. Chem.* **2013**, *288*, 4715–4722. [CrossRef]
8. Vandavasi, V.G.; Weiss, K.L.; Cooper, J.B.; Erskine, P.T.; Tomanicek, S.J.; Ostermann, A.; Schrader, T.E.; Ginell, S.L.; Coates, L. Exploring the Mechanism of β-Lactam Ring Protonation in the Class A β-lactamase Acylation Mechanism Using Neutron and X-ray Crystallography. *J. Med. Chem.* **2016**, *59*, 474–479. [CrossRef]
9. Ibuka, A.; Taguchi, A.; Ishiguro, M.; Fushinobu, S.; Ishii, Y.; Kamitori, S.; Okuyama, K.; Yamaguchi, K.; Konno, M.; Matsuzawa, H. Crystal structure of the E166A mutant of extended-spectrum β-lactamase toho-1 at 1.8 Å resolution11Edited by R. Huber. *J. Mol. Biol.* **1999**, *285*, 2079–2087. [CrossRef]
10. Ibuka, A.S.; Ishii, Y.; Galleni, M.; Ishiguro, M.; Yamaguchi, K.; Frère, J.-M.; Matsuzawa, H.; Sakai, H. Crystal Structure of Extended-Spectrum β-Lactamase Toho-1: Insights into the Molecular Mechanism for Catalytic Reaction and Substrate Specificity Expansion. *Biochemistry* **2003**, *42*, 10634–10643. [CrossRef]

11. Shimizu-Ibuka, A.; Matsuzawa, H.; Sakai, H. An Engineered Disulfide Bond between Residues 69 and 238 in Extended-Spectrum β-Lactamase Toho-1 Reduces Its Activity toward Third-Generation Cephalosporins. *Biochemistry* **2004**, *43*, 15737–15745. [CrossRef] [PubMed]

12. Kelly, J.A.; Dideberg, O.; Charlier, P.; Wery, J.P.; Libert, M.; Moews, P.C.; Knox, J.R.; Duez, C.; Fraipont, C.; Joris, B.; et al. On the origin of bacterial resistance to penicillin: Comparison of a beta-lactamase and a penicillin target. *Science* **1986**, *231*, 1429. [CrossRef] [PubMed]

13. Urbach, C.; Evrard, C.; Pudzaitis, V.; Fastrez, J.; Soumillion, P.; Declercq, J.-P. Structure of PBP-A from Thermosynechococcus elongatus, a Penicillin-Binding Protein Closely Related to Class A β-Lactamases. *J. Mol. Biol.* **2009**, *386*, 109–120. [CrossRef]

14. Pratt, R.F. β-Lactamases: Why and How. *J. Med. Chem.* **2016**, *59*, 8207–8220. [CrossRef] [PubMed]

15. Silvaggi, N.R.; Josephine, H.R.; Kuzin, A.P.; Nagarajan, R.; Pratt, R.F.; Kelly, J.A. Crystal Structures of Complexes between the R61 DD-peptidase and Peptidoglycan-mimetic β-Lactams: A Non-covalent Complex with a "Perfect Penicillin" This article is dedicated to the memory of Professor Jean-Marie Ghuysen of the Université de Liège, Belgium, in recognition of his decades of contributions towards our understanding of and ability to combat bacterial infections. *J. Mol. Biol.* **2005**, *345*, 521–533. [CrossRef] [PubMed]

16. Wilkin, J.M.; Dubus, A.; Joris, B.; Frère, J.M. The mechanism of action of DD-peptidases: The role of Threonine-299 and -301 in the Streptomyces R61 DD-peptidase. *Biochem. J* **1994**, *301*, 477. [CrossRef] [PubMed]

17. Tipper, D.J.; Strominger, J.L. Mechanism of action of penicillins: A proposal based on their structural similarity to acyl-D-alanyl-D-alanine. *Proc. Natl. Acad. Sci. USA* **1965**, *54*, 1133. [CrossRef]

18. Cheng, F.; Liu, C.; Jiang, J.; Lu, W.; Li, W.; Liu, G.; Zhou, W.; Huang, J.; Tang, Y. Prediction of Drug-Target Interactions and Drug Repositioning via Network-Based Inference. *PLoS Comp. Biol.* **2012**, *8*, e1002503. [CrossRef]

19. Öztürk, H.; Ozkirimli, E.; Özgür, A. Classification of Beta-Lactamases and Penicillin Binding Proteins Using Ligand-Centric Network Models. *PLoS ONE* **2015**, *10*, e0117874. [CrossRef]

20. Urbach, C.; Fastrez, J.; Soumillion, P. A New Family of Cyanobacterial Penicillin-binding Proteins: A MISSING LINK IN THE EVOLUTION OF CLASS A β-LACTAMASES. *J. Biol. Chem.* **2008**, *283*, 32516–32526. [CrossRef]

21. Hargis, J.C.; White, J.K.; Chen, Y.; Woodcock, H.L. Can Molecular Dynamics and QM/MM Solve the Penicillin Binding Protein Protonation Puzzle? *J. Chem. Inf. Model.* **2014**, *54*, 1412–1424. [CrossRef] [PubMed]

22. Gherman, B.F.; Goldberg, S.D.; Cornish, V.W.; Friesner, R.A. Mixed Quantum Mechanical/Molecular Mechanical (QM/MM) Study of the Deacylation Reaction in a Penicillin Binding Protein (PBP) versus in a Class C β-Lactamase. *J. Am. Chem. Soc.* **2004**, *126*, 7652–7664. [CrossRef] [PubMed]

23. Pratt, R.F. Functional evolution of the serine β-lactamase active site. *J. Chem. Soc. Perk. Trans. 2* **2002**, 851–861. [CrossRef]

24. Noé, F.; Schütte, C.; Vanden-Eijnden, E.; Reich, L.; Weikl, T.R. Constructing the equilibrium ensemble of folding pathways from short off-equilibrium simulations. *Proc. Natl. Acad. Sci. USA* **2009**, *106*, 19011. [CrossRef] [PubMed]

25. Noé, F.; Wu, H.; Prinz, J.-H.; Plattner, N. Projected and hidden Markov models for calculating kinetics and metastable states of complex molecules. *J. Chem. Phys.* **2013**, *139*, 184114. [CrossRef]

26. Wang, F.; Shen, L.; Zhou, H.; Wang, S.; Wang, X.; Tao, P. Machine Learning Classification Model for Functional Binding Modes of TEM-1 β-Lactamase. *Front. Mol. Biosci.* **2019**, *6*, 47. [CrossRef]

27. Best, R.B.; Zhu, X.; Shim, J.; Lopes, P.E.; Mittal, J.; Feig, M.; MacKerell, A.D., Jr. Optimization of the additive CHARMM all-atom protein force field targeting improved sampling of the backbone φ, ψ and side-chain χ1 and χ2 dihedral angles. *J. Chem. Theory Comput.* **2012**, *8*, 3257–3273. [CrossRef]

28. Vanommeslaeghe, K.; Hatcher, E.; Acharya, C.; Kundu, S.; Zhong, S.; Shim, J.; Darian, E.; Guvench, O.; Lopes, P.; Vorobyov, I.; et al. CHARMM general force field: A force field for drug-like molecules compatible with the CHARMM all-atom additive biological force fields. *J. Comput. Chem.* **2010**, *31*, 671–690. [CrossRef]

29. Yu, W.; He, X.; Vanommeslaeghe, K.; MacKerell, A.D., Jr. Extension of the CHARMM general force field to sulfonyl-containing compounds and its utility in biomolecular simulations. *J. Comput. Chem.* **2012**, *33*, 2451–2468. [CrossRef]

30. Jorgensen, W.L.; Chandrasekhar, J.; Madura, J.D.; Impey, R.W.; Klein, M.L. Comparison of simple potential functions for simulating liquid water. *J. Chem. Phys.* **1983**, *79*, 926–935. [CrossRef]

31. Neria, E.; Fischer, S.; Karplus, M. Simulation of activation free energies in molecular systems. *J. Chem. Phys.* **1996**, *105*, 1902–1921. [CrossRef]

32. Ryckaert, J.-P.; Ciccotti, G.; Berendsen, H.J.C. Numerical integration of the cartesian equations of motion of a system with constraints: Molecular dynamics of n-alkanes. *J. Comput. Phys.* **1977**, *23*, 327–341. [CrossRef]

33. Darden, T.; York, D.; Pedersen, L. Particle mesh Ewald: An N log (N) method for Ewald sums in large systems. *J. Chem. Phys.* **1993**, *98*, 10089–10092. [CrossRef]

34. Friedrichs, M.S.; Eastman, P.; Vaidyanathan, V.; Houston, M.; Legrand, S.; Beberg, A.L.; Ensign, D.L.; Bruns, C.M.; Pande, V.S. Accelerating molecular dynamic simulation on graphics processing units. *J. Comput. Chem.* **2009**, *30*, 864–872. [CrossRef] [PubMed]

35. Eastman, P.; Pande, V. OpenMM: A hardware-independent framework for molecular simulations. *Comput. Sci. Eng.* **2010**, *12*, 34–39. [CrossRef] [PubMed]

36. Eastman, P.; Swails, J.; Chodera, J.D.; McGibbon, R.T.; Zhao, Y.; Beauchamp, K.A.; Wang, L.-P.; Simmonett, A.C.; Harrigan, M.P.; Stern, C.D. OpenMM 7: Rapid development of high performance algorithms for molecular dynamics. *PLoS Comp. Biol.* **2017**, *13*, e1005659. [CrossRef] [PubMed]

37. Wang, F.; Zhou, H.; Olademehin, O.P.; Kim, S.J.; Tao, P. Insights into Key Interactions between Vancomycin and Bacterial Cell Wall Structures. *ACS Omega* **2018**, *3*, 37–45. [CrossRef]

38. Jolliffe, I. Principal component analysis. In *International Encyclopedia of Statistical Science*; Springer: New York, NY, USA, 2011; pp. 1094–1096.

39. Levy, R.M.; Srinivasan, A.R.; Olson, W.K.; McCammon, J.A. Quasi-harmonic method for studying very low frequency modes in proteins. *Biopolymers* **1984**, *23*, 1099–1112. [CrossRef]

40. Zhou, H.; Wang, F.; Tao, P. t-Distributed Stochastic Neighbor Embedding Method with the Least Information Loss for Macromolecular Simulations. *J. Chem. Theory Comput.* **2018**, *14*, 5499–5510. [CrossRef]

41. Schwantes, C.R.; Pande, V.S. Improvements in Markov State Model Construction Reveal Many Non-Native Interactions in the Folding of NTL9. *J. Chem. Theory Comput.* **2013**, *9*, 2000–2009. [CrossRef]

42. Pérez-Hernández, G.; Paul, F.; Giorgino, T.; De Fabritiis, G.; Noé, F. Identification of slow molecular order parameters for Markov model construction. *J. Chem. Phys.* **2013**, *139*, 015102. [CrossRef] [PubMed]

43. Naritomi, Y.; Fuchigami, S. Slow dynamics in protein fluctuations revealed by time-structure based independent component analysis: The case of domain motions. *J. Chem. Phys.* **2011**, *134*, 065101. [CrossRef] [PubMed]

44. Pinamonti, G.; Zhao, J.; Condon, D.E.; Paul, F.; Noè, F.; Turner, D.H.; Bussi, G. Predicting the Kinetics of RNA Oligonucleotides Using Markov State Models. *J. Chem. Theory Comput.* **2017**, *13*, 926–934. [CrossRef] [PubMed]

45. Plattner, N.; Noé, F. Protein conformational plasticity and complex ligand-binding kinetics explored by atomistic simulations and Markov models. *Nat. Commun.* **2015**, *6*, 7653. [CrossRef]

46. Chodera, J.D.; Elms, P.; Noé, F.; Keller, B.; Kaiser, C.M.; Ewall-Wice, A.; Marqusee, S.; Bustamante, C.; Hinrichs, N.S. Bayesian hidden Markov model analysis of single-molecule force spectroscopy: Characterizing kinetics under measurement uncertainty. *arXiv* **2011**, arXiv:1108.1430.

47. Trendelkamp-Schroer, B.; Wu, H.; Paul, F.; Noé, F. Estimation and uncertainty of reversible Markov models. *J. Chem. Phys.* **2015**, *143*, 174101. [CrossRef]

48. Vanden-Eijnden, E. Transition Path Theory. In *An Introduction to Markov State Models and Their Application to Long Timescale Molecular Simulation*; Bowman, G.R., Pande, V.S., Noé, F., Eds.; Springer: Dordrecht, The Netherlands, 2014; pp. 91–100. [CrossRef]

49. David, C.C.; Jacobs, D.J. Principal Component Analysis: A Method for Determining the Essential Dynamics of Proteins. In *Protein Dynamics: Methods and Protocols*; Livesay, D.R., Ed.; Humana Press: Totowa, NJ, USA, 2014; pp. 193–226. [CrossRef]

50. Zhou, Y.; Yu, Z.-G.; Anh, V. Cluster protein structures using recurrence quantification analysis on coordinates of alpha-carbon atoms of proteins. *Phys. Lett. A* **2007**, *368*, 314–319. [CrossRef]

51. Webber, C.L., Jr.; Giuliani, A.; Zbilut, J.P.; Colosimo, A. Elucidating protein secondary structures using alpha-carbon recurrence quantifications. *Proteins Struct. Funct. Bioinform.* **2001**, *44*, 292–303. [CrossRef]

52. Scherer, M.K.; Trendelkamp-Schroer, B.; Paul, F.; Pérez-Hernández, G.; Hoffmann, M.; Plattner, N.; Wehmeyer, C.; Prinz, J.-H.; Noé, F. PyEMMA 2: A Software Package for Estimation, Validation, and Analysis of Markov Models. *J. Chem. Theory Comput.* **2015**, *11*, 5525–5542. [CrossRef]

53. Rabiner, L.R. A tutorial on hidden Markov models and selected applications in speech recognition. *Proc. IEEE* **1989**, *77*, 257–286. [CrossRef]

54. Baum, L.E.; Petrie, T.; Soules, G.; Weiss, N. A Maximization Technique Occurring in the Statistical Analysis of Probabilistic Functions of Markov Chains. *Ann. Math. Stat.* **1970**, *41*, 164–171. [CrossRef]

55. Russell, R.B.; Barton, G.J. Multiple protein sequence alignment from tertiary structure comparison: Assignment of global and residue confidence levels. *Proteins Struct. Funct. Bioinform.* **1992**, *14*, 309–323. [CrossRef] [PubMed]

56. Donoghue, P.; Luthey-Schulten, Z. On the Evolution of Structure in Aminoacyl-tRNA Synthetases. *Microbiol. Mol. Biol. Rev.* **2003**, *67*, 550. [CrossRef] [PubMed]

57. Hermann, J.C.; Hensen, C.; Ridder, L.; Mulholland, A.J.; Höltje, H.-D. Mechanisms of Antibiotic Resistance: QM/MM Modeling of the Acylation Reaction of a Class A β-Lactamase with Benzylpenicillin. *J. Am. Chem. Soc.* **2005**, *127*, 4454–4465. [CrossRef]

58. Castillo, R.; Silla, E.; Tuñón, I. Role of Protein Flexibility in Enzymatic Catalysis: Quantum Mechanical–Molecular Mechanical Study of the Deacylation Reaction in Class A β-Lactamases. *J. Am. Chem. Soc.* **2002**, *124*, 1809–1816. [CrossRef] [PubMed]

59. Swarén, P.; Maveyraud, L.; Raquet, X.; Cabantous, S.; Duez, C.; Pédelacq, J.-D.; Mariotte-Boyer, S.; Mourey, L.; Labia, R.; Nicolas-Chanoine, M.-H. X-ray analysis of the NMC-A β-lactamase at 1.64-Å resolution, a class A carbapenemase with broad substrate specificity. *J. Biol. Chem.* **1998**, *273*, 26714–26721. [CrossRef] [PubMed]

60. Díaz, N.; Suárez, D.; Sordo, T.L.; Merz, K.M. Acylation of class A β-lactamases by penicillins: A theoretical examination of the role of serine 130 and the β-lactam carboxylate group. *J Phys. Chem. B* **2001**, *105*, 11302–11313. [CrossRef]

61. Noé, F.; Horenko, I.; Schütte, C.; Smith, J.C. Hierarchical analysis of conformational dynamics in biomolecules: Transition networks of metastable states. *J. Chem. Phys.* **2007**, *126*, 155102. [CrossRef]

62. Bös, F.; Pleiss, J. Multiple Molecular Dynamics Simulations of TEM β-Lactamase: Dynamics and Water Binding of the Ω-Loop. *Biophys. J.* **2009**, *97*, 2550–2558. [CrossRef]

63. Fisette, O.; Morin, S.; Savard, P.-Y.; Lagüe, P.; Gagné, S.M. TEM-1 Backbone Dynamics—Insights from Combined Molecular Dynamics and Nuclear Magnetic Resonance. *Biophys. J.* **2010**, *98*, 637–645. [CrossRef]

64. Horn, J.R.; Shoichet, B.K. Allosteric Inhibition through Core Disruption. *J. Mol. Biol.* **2004**, *336*, 1283–1291. [CrossRef] [PubMed]

65. Matagne, A.; Lamotte-Brasseur, J.; FrÈRe, J.-M. Catalytic properties of class A β-lactamases: Efficiency and diversity. *Biochem. J.* **1998**, *330*, 581. [CrossRef] [PubMed]

66. Imtiaz, U.; Billings, E.; Knox, J.R.; Manavathu, E.K.; Lerner, S.A.; Mobashery, S. Inactivation of class A. beta.-lactamases by clavulanic acid: The role of arginine-244 in a proposed nonconcerted sequence of events. *J. Am. Chem. Soc.* **1993**, *115*, 4435–4442. [CrossRef]

67. Shimizu-Ibuka, A.; Oishi, M.; Yamada, S.; Ishii, Y.; Mura, K.; Sakai, H.; Matsuzawa, H. Roles of Residues Cys69, Asn104, Phe160, Gly232, Ser237, and Asp240 in Extended-Spectrum β-Lactamase Toho-1. *Antimicrob. Agents Chemother.* **2011**, *55*, 284. [CrossRef]

68. Langan, P.S.; Vandavasi, V.G.; Cooper, S.J.; Weiss, K.L.; Ginell, S.L.; Parks, J.M.; Coates, L. Substrate Binding Induces Conformational Changes in a Class A β-lactamase That Prime It for Catalysis. *ACS Catalysis* **2018**, *8*, 2428–2437. [CrossRef]

69. Doucet, N.; Savard, P.-Y.; Pelletier, J.N.; Gagné, S.M. NMR Investigation of Tyr105 Mutants in TEM-1 β-Lactamase: DYNAMICS ARE CORRELATED WITH FUNCTION. *J. Biol. Chem.* **2007**, *282*, 21448–21459. [CrossRef]

70. McDonough, M.A.; Anderson, J.W.; Silvaggi, N.R.; Pratt, R.F.; Knox, J.R.; Kelly, J.A. Structures of Two Kinetic Intermediates Reveal Species Specificity of Penicillin-binding Proteins. *J. Mol. Biol.* **2002**, *322*, 111–122. [CrossRef]

71. Silvaggi, N.R.; Anderson, J.W.; Brinsmade, S.R.; Pratt, R.F.; Kelly, J.A. The Crystal Structure of Phosphonate-Inhibited d-Ala-d-Ala Peptidase Reveals an Analogue of a Tetrahedral Transition State. *Biochemistry* **2003**, *42*, 1199–1208. [CrossRef]

Article

Non-Linear Dynamics Analysis of Protein Sequences. *Application to CYP450*

Xavier F. Cadet [1,2], Reda Dehak [2], Sang Peter Chin [3] and Miloud Bessafi [4,*

[1] PEACCEL, Protein Engineering Accelerator, 6 square Albin Cachot, box 42, 75013 Paris, France
[2] LSE laboratory, EPITA, Paris 94276, France
[3] Learning Intelligence Signal Processing Group, Department of Computer Science, Boston University, Boston, MA 02215, USA
[4] LE2P-Energy Lab, Laboratory of Energy, Electronics and Processes EA 4079, Faculty of Sciences and Technology, University of La Reunion, 97444 St Denis CEDEX, France
* Correspondence: miloud.bessafi@univ-reunion.fr

Received: 30 June 2019; Accepted: 29 August 2019; Published: 31 August 2019

Abstract: The nature of changes involved in crossed-sequence scale and inner-sequence scale is very challenging in protein biology. This study is a new attempt to assess with a phenomenological approach the non-stationary and nonlinear fluctuation of changes encountered in protein sequence. We have computed fluctuations from an encoded amino acid index dataset using cumulative sum technique and extracted the departure from the linear trend found in each protein sequence. For inner-sequence analysis, we found that the fluctuations of changes statistically follow a $-5/3$ Kolmogorov power and behave like an incremental Brownian process. The pattern of the changes in the inner sequence seems to be monofractal in essence and to be bounded between Hurst exponent [1/3,1/2] range, which respectively corresponds to the Kolmogorov and Brownian monofractal process. In addition, the changes in the inner sequence exhibit moderate complexity and chaos, which seems to be coherent with the monofractal and stochastic process highlighted previously in the study. The crossed-sequence changes analysis was achieved using an external parameter, which is the activity available for each protein sequence, and some results obtained for the inner sequence, specifically the drift and Kolmogorov complexity spectrum. We found a significant linear relationship between activity changes and drift changes, and also between activity and Kolmogorov complexity. An analysis of the mean square displacement of trajectories in the bivariate space (drift, activity) and (Kolmogorov complexity spectrum, activity) seems to present a superdiffusive law with a 1.6 power law value.

Keywords: power law; Brownian process; Kolmogorov complexity; entropy; chaos; monofractal; non-linear; cumulative sum; sequence analysis; protein engineering

---

1. Introduction

From the information viewpoint, a protein sequence can be considered as a distribution of successive symbols extracted with a rule from a dictionary. Conceptually, it means that the protein sequence is simply encoded to a set of symbol combinations. Moreover, the number of the symbols used is usually very small in comparison to the length of the protein sequence. Consequently, there is a huge variety of combinations of symbols to encode a protein sequence in the real world. It is well-known that the molecular mechanism (stability, structure function, disorder) is often triggered by complex interactions [1–3]. Like the emerged part of an iceberg, the intricated symbol set of an encoded protein sequence can be seen as a footprint of a wide range of covert biochemical interactions within the protein. Then, there are numerous encoder models that try to reflect the reality accurately using a conversion rule related to physicochemical and biochemical properties [4–6]. Beyond the symbol combination and arrangement of the protein sequence, understanding the nature and the organization of the symbols is very challenging in protein biology. Therefore, analyzing the encoded protein sequence by means of nonlinear analysis can provide some insights about the dynamics of the changes within the dataset. Searching for similarities between encoded protein sequences in a dataset is one of the important advantages of morphological analysis of protein sequences. There are many approaches to extract groups, which are conceptually based on a clustering method of global or local information about the protein sequence [7–13]. The prediction of disorder of the protein sequence is often related to the ability to track the degree of randomness, the stochasticity, and the complexity embedded in the whole encoded dataset. There are studies which focus on randomness, chaos, long-range interaction between sequences for classification, and predictability. For example, Yu et al. [14] have made a comparative study of structure and intrinsic disorder between 10,000 natural and random protein sequences and found that natural sequences have more long disordered regions than random sequences. In addition, Gök et al. [5] have used the Lyapunov exponent and test four classifier algorithms (Bayesian network, Naïve Bayes, k-means, and SVM) to identify the disordered protein regions. Long short-term memory (LSTM) recurrent neural networks is a deep learning algorithm that has gained some interest for tracking the long-range interactions between sequences [1,15]. These studies reveal that there is potential information about degree of randomness, disorder, and stochasticity in protein sequences and beyond some degree of predictability. It means that the protein sequence exhibits some order within disorder and changes are not a likelihood for this set of symbols. To find out what kind of information and properties of disorder or complexity we are able to extract from protein sequences, we propose to scan the changes inside the protein sequences and between sequences using a multidisciplinary approach. It means that we intend, at the same time, to use tools from information theory field (entropy of information, Kolmogorov complexity), physical theory (chaos, fractional Brownian processes, drift-diffusion processes), and signal processing (multifractality, Fourier analysis). To our knowledge, the use of multidisciplinary tools to analyze the dynamics of the changes within a protein sequence and between sequences is new. As mentioned previously, the encoded protein sequence contains successive numerical values and can also be considered as a time series. The aim of this paper is to encompass the variability of the inner changes hidden behind the encoded protein sequence using nonlinear tools, and to assess the predictability of the underlying non-stationary protein sequence activity.

The study is organized as follows. Section 2 presents the experimental dataset and the encoded protein sequence. Section 3 describes the algorithm used to analyze the time series (i) entropy and chaos, (ii) Kolmogorov complexity and Turing machine, (iii) law-scaling and stochastic process, and (iv) surrogated and shuffled data. Finally, Section 4 includes both presentation of the results obtained and discussion. The concluding remarks are given in Section 5.

2. Experimental Dataset

To facilitate the understanding of readers outside the realm of life sciences, we will provide a brief definition of a polypeptide/protein sequence. A protein sequence is a chain made of residues of amino acids. Twenty amino acids are the basic building blocks for proteins. We will provide an application example as well.

2.1. Alphabetical Dictionary

Each amino acid is represented by a letter corresponding to the one-letter code for an amino acid. The global sequence has a biological meaning. A single variation in the sequence could have a huge impact on the activity of the protein. An example of a protein sequence (Cytochrome P450) is given below:

MTIKEMPQPKTFGELKNLPLLNTDKPVQALMKIADELGEIFKFEAPGRVTRYLSSQRLIKEACDES
RFDKNLSQALKFVRDFAGDGLATSWTHEKNWKKAHNILLPSFSQQAMKGYHAMMVDIATQLI
QKWSRLNPNEEIDVADDMTRLTLDTIGLCGFNYRFNSFYRDSQHPFITSMLRALKEAMNQSKRL
LRLWPTAPAFSLYAKEDTVLGGEYPLEKGDELMVLIPQLHRDKTIWGDDVEEFRPERFENPSAIPQ
HAFKPFGNGQRACIGQQFALHEATLVLGMILKYFTLIDHENYELDIKQTLTLKPGDFHISVQSRH
QEAIHADVQAAE

2.2. An Application Example: Cytochrome P450

Cytochrome P450 is a protein, i.e., a polypeptidic sequence of 464 or 466 amino acids. It is used to generate products of significant medical and industrial importance. Three parental cytochromes P450, i.e., CYP102A1(P1), CYP102A2(P2), and CYP102A3(P3) were used to generate 242 chimeric sequences of cytochrome P450 [16]. Further, 242 thermostable protein sequences were created by recombination of stabilizing fragments. For each variant, the thermostability (defined herewith as: Activity) was analyzed by the measurement of the T_{50}, T_{50} being the temperature at which 50% of the protein was irreversibly denatured after incubation for 10 min. The result is a decrease in activity. Activity ranges from 39.2 °C to 64.48 °C. Chimeras are written according to fragment composition: 23121321 represents a protein that inherits the first fragment from parent P2, the second from P3, the third from P1, and so on.

3. Methodology

In this study, the questions are: "Can statistical, nonlinear, and complexity analysis give us some information about the pattern in a protein sequence and its changes along the sequence and also the next, or other sequences? Can we group sequences according to their activity but also their morphological pattern?". To assess the ability of the statistical chaos and complexity tools, we have transformed each protein sequence into numerical or binary time series according to the need of the use of the tool.

First of all, there exist different conversion tables to transform protein residues (letters) to numerical sequences. We have used the freely available one, namely AA index database [17,18]. This database contains a huge number of ascribed numerical values for each protein residue. There are 566 numerical values, which are for each index in the sequence univocally in correspondence with physicochemical and biochemical properties of the residues. In this case, we have selected the index 532 in the dataset, which allows us to rank and encode 20 standard amino acids.

3.1. Entropy and Chaos

Entropy is a concept that was first discovered in physics. Nevertheless, this concept is also encountered in other fields and especially in the theory of information. In 1948, Shannon [19] formalized the concept of entropy of the information H of a string of length N, which contains Q repeated symbols $S = \{s_1, s_2, \ldots, s_Q\}$. H is shown by the well-known formula:

$$H = -\sum_{i=1}^{Q} \hat{p}_i log \hat{p}_i \tag{1}$$

where $\hat{p}_i = \frac{N_{s_i}}{N}$.

N_{s_i} is the number of appearances of the symbol s_i in the string of length N. Thus, p_i is the probability of occurrence within the range value $]0\ 1]$. As we suppose that all Q symbols exist in the string, the probability 0 is excluded. The minus sign is to ensure a positive value of the entropy H as the logarithm is always negative. H is a global measure of the total amount of information in an entire probability distribution contained in a sequence.

Another measure of entropy is the sample entropy [20]. Let us consider a set of N symbols $s_{i,k}$ in a sequence S_i chosen among M sequences in the dataset. From the sequence S_i we extract two subsets of m symbols $S_{i,p}^m = \{s_{i,p},\ s_{i,p+1}, \ldots, s_{i,p+m}\}$ and $S_{iq}^m = \{s_{i,q},\ s_{i,q+1}, \ldots, s_{i,q+m}\}$ where $p \neq q$.

The parameters p and q correspond to the index position of the first symbol of respectively the subset $S_{i,p}^m$ and $S_{i,q}^m$ within the sequence S_i. The sample entropy (*SampEn*) of the sequence S_i is defined as $SampEn(m, r, N)_i = -log\left(\frac{A_i}{B_i}\right)$, where A_i is the number of pair-wise subset symbols $\left(s_{ip}^{m+1}, s_{iq}^{m+1}\right)$ of length $m + 1$ with a distance $d\left(s_{ip}^{m+1}, s_{iq}^{m+1}\right) < r$ while B_i is the number of pair-wise subset symbols $\left(s_{ip}^m, s_{iq}^m\right)$ of length m with a distance $d\left(s_{ip}^m, s_{iq}^m\right) < r$. The r is a threshold value of similarity between the pair-wise subset symbols $\left(s_{ip}^m, s_{iq}^m\right)$. In our study, the sequence is a set of numbers. Then, the distance $d\left(s_{ip}^m, s_{iq}^m\right)$ is a Euclidian distance and the tolerance value threshold value r is chosen between 0.1 and 0.2 of the standard deviation of the sequence S_i [20]. Moreover, the embedding dimension m is usually taken to be 2. Finally, the sample entropy is a positive value, which can be 0 for a regular sequence and roughly 2.2 or 2.3 for a strongly irregular sequence. The sample entropy is a measure of the regularity within a sequence.

In addition, sometimes an irregularity pattern in a time series could be related to the chaos process within a sequence. The largest Lyapunov exponent is the most common parameter used to characterize chaos in a dynamical system. The sign and the value of this parameter give an indication of the response of a system to amplify, damp, or oscillate a small perturbation. In our case, it means that if the largest Lyapunov exponent is (i) positive, then the process is chaotic, (ii) close to zero, then the process is periodic or quasi-periodic, and finally (iii) negative, the process is damping and has an attractor. In our study, to achieve the search for chaos pattern in a sequence S_i, we have used Wolf's algorithm [21] to compute the Lyapunov exponent spectrum and the largest Lyapunov exponent (*LLE*).

3.2. Kolmogorov Complexity and Turing Machine

Let us assume we have a set of M sequences $S = \{S_1, S_2, \ldots, S_M\}$. Then, we suppose that we have for each sequence i of string S_i, a set of N values defined as $S_i = \{p_i^1, p_i^2, \ldots, p_i^N\}$. To assess disorder within a sequence, we use the Kolmogorov complexity method [22]. This method is based on the concept of Turing machine and the mathematical expression of the algorithmic complexity can be written $K_T(s) = min\{|p|, T(p) = s\}$. This states that the algorithmic complexity of a string s is the shortest program p computed with a Turing's machine T to gather output s [23,24]. To compute the Kolmogorov complexity (*KC*), there are three processes: (i) Convert the sequence S_i to binary sequence B_i using a threshold method, (ii) compress the sequence B_i with Lempel-Ziv compressor to a compressed sequence C_i, and (iii) compute and normalize the Kolmogorov complexity number associated with the original sequence S_i.

Binarizing the sequence S_i is based on the particular value used as threshold value p_i^T to assign each number p_i^k in the sequence S_i with the value of 0 if p_i^k is less than the threshold value p_i^T, or conversely assigned with the value of 1 if p_i^k exceeds the threshold value p_i^T. The mathematical expression of the binary value of the number p_i^k in the sequence S_i is:

$$B_i^k\Big|_{\substack{i = \{1, 2, \ldots, M\} \\ k = \{1, 2, \ldots, N\}}} = \begin{cases} 0 & if\ p_i^k < p_i^T \\ or & \\ 1 & if\ p_i^k \geq p_i^T \end{cases} \tag{2}$$

where p_i^T is a threshold value of sequence S_i.

Usually, the mean of the set $\{p_i^1, p_i^2, \ldots, p_i^N\}$ is used as a threshold value of the sequence S_i. Nevertheless, we will take into account the amplitude of the numbers p_i^k to compute the optimum threshold value $p_i^T{}_{opt}$ associated with the sequence S_i. Thus, we introduce the Kolmogorov complexity spectrum (*KCS*), which is an iterative procedure to compute the Kolmogorov complexity for various threshold values within the range values p_i^k of the sequence S_i [25]. The encoding number to binary value is presented as:

$$B_i^k\big|_m = \begin{cases} 0 & if \ p_i^k < p_i^{T_m} \\ & or \\ 1 & if \ p_i^k \geq p_i^{T_m} \end{cases} \qquad m = 1, 2, \ldots K \qquad (3)$$

where $p_i^{T_L} = min_k\big(\{p_i^k\}\big) + m\left\{\dfrac{max_k(\{p_i^k\}) - min_k(\{p_i^k\})}{K-1}\right\}$.

Thus, for each sequence S_i, the Kolmogorov complexity spectrum is a set of K Kolmogorov complexity values $KC_i^K = \{KC_i^1, \ KC_i^2, \ \ldots, \ KC_i^K\}$. The optimum threshold $p_i^T{}_{opt}$ is chosen among the set of threshold values $\{p_i^{T_1}, \ p_i^{T_2}, \ \ldots, \ p_i^{T_K}\}$ using the condition $p_i^T{}_{opt} = \{p_i^{T_j} \ \big| \ KC_i^j = max_k(KC_i^k)\}$.

The compression method used in this study is the Lempel-Ziv compressor [26]. This is an iterative search in the binary series B_i of the overall possible subset sequences, which are different from each other. The result is a compressed sequence C_i. If $|C_i|$ represents the length of the compressed binary sequence C_i, then Kolmogorov complexity KC_i associated with the sequence S_i is:

$$KC_i = |C_i| log_2 N / N. \qquad (4)$$

The term $log_2 N / N$ in the expression of KC_i insures the normalization of the Kolmogorov complexity.

3.3. Law-Scaling and Stochastic Process

As previously mentioned, a sequence is defined as a set of alphabetic letters, which could be converted to other symbols (numerical, binary, etc.). Nevertheless, the changes of symbols along the chain are usually related to the real world of biochemical activities along the protein sequence. The question is *"Do those changes present a regular or irregular pattern within a sequence which can provide some information about an underlying dynamic in a sequence?"* First, we have to define the changes in a sequence i of pairwise symbols separated by a distance, namely an increment of position. Let us assume d is the increment pairwise symbols and the quantity $\Delta p_{d_i} = \big|p_i^j - p_i^k\big|_{d=|k-j|}$ is the magnitude of changes of the pairwise symbols separated by an increment of d. We define the structure function $S_{q_i}(d)$ for a sequence i defined by the expression $S_{q_i}(d) = \frac{1}{N_{d_i}} \sum_{m=1}^{N_{d_i}} |p_i^j - p_i^k|_{d=|k-j|}^q$ where N_{d_i} is the number of pairwise symbols separated with a distance d. By extension, this function can also be used to track the existence of scaling law in the data $S_{q_i}(d) \propto d^{\xi(q)}$. $\xi(q)$ is the generalized Hurst exponent, which is indicative of the nature of pairwise symbol changes and the stochasticity of processes like long-term memories, Brownian motion, self-similarity pattern [27]. The probability function (PDFs) of the distribution of the normalized changes of pair-wise symbols $\Delta p_{d_i} / \sigma\big(\Delta p_{d_i}\big)$ within a sequence i can be computed to analyze the normality of the changes in a sequence. Additionally, kurtosis or flatness is another measure of the normality of the changes of the pairwise symbols. For sequence i, the kurtosis $F_i = S_{4i}(d) / (S_{2i}(d))^2$. The terms $S_{4i}(d)$ and $S_{2i}(d)$ are, respectively, the fourth- and second-order moment of the pairwise distribution.

3.4. Surrogated and Shuffled Data

The methods to surrogate and shuffle the data are very popular tools to assess the existence of nonlinearities and the scaling properties of a process. Both algorithms are based on the generation of randomized synthetic data using specific constraint rule to generate the synthetic data. Surrogated data used in this study are the iterative amplitude-adjusted Fourier transform (IAAFT). This method preserves the statistical properties of the original data but randomizes the phase spectrum of the Fourier transform of the original data. The synthetic data generated with this method lead to removing nonlinearities in the original data. Shuffled data are obtained by a random permutation between values of the original data. This method is a bootstrapping algorithm without repetition of the indices' permutation. Variants of the protein (synthetic sequences) are obtained by variation of any position in the sequence and not by variation of the fragments constitutive of the protein (described in the Section 2.2 "An application example: Cytochrome P450"). The data obtained are a set of values that do not exhibit any linear correlation in the synthetic data and preserve the amplitude distribution. For more information about these two algorithms, the reader can refer to the review of Schreiber and Schmitz [28].

4. Normalized Detrended Cumulative Sum (NDCS) Method

Fluctuations or changes along the protein sequence are of interest in this study but we need to show how we extract this information from the original data. Cumulative sum is a sequential method that is widely used to detect changes in a time series and to track the self-similarity in a dataset [29]. In this study, we have applied this algorithm for each sequence and generated a new sequence of fluctuations defined as a departure from the linear trend. Within the 242 protein sequences of a length of 466 for each one, each index in a sequence is originally labelled with an alphabetical letter. There are 20 letters used (A, C, D, E, F, G, H, I, K, L, M, N, P, Q, R, S, T, V, W, and Y) corresponding to the one-letter code for amino acid. In this study, the D PRIFT index is chosen from the AA index catalog to convert the alphabetical symbols to numerical values [30]. It allows us to distinguish each of the 20 amino acid residues by a unique value related to its hydrophobicity property. The encoding process, which converts the original alphabetical letters to numerical values within the [−5.68 6.81] range, is shown in Table 1.

Table 1. Conversion rule of protein sequence of AA index 532—D PRIFT index [30].

AA Index 532 D PRIFT Index (Cornette et al. 1987)																			
Letter	A	C	D	E	F	G	H	I	K	L	M	N	P	Q	R	S	T	V	W	Y
Value Index	−5.68	−5.62	−5.30	−4.47	−3.99	−3.86	−1.94	−1.92	−1.28	0.96	0.62	0.21	0.75	3.34	4.54	4.76	5.06	5.39	5.54	6.81

We are aware that this description by their hydrophobicity values is oversimplified and does not account (i) for many other properties of amino acids that are well known to strongly affect pattern changes in protein sequences along families, such as volume, aromaticity, and different charge states for the same amino acid in distinct positions or, (ii) for the fact that the exposure of continuous amino acids sequences to solvent or their occlusion in protein cores is a fundamental requirement for proteins to fold in functional arrangements, giving importance to hydrophobic and polar amino acids and their distribution. However, whatever the choice among all the possible amino acid indexes that are able to distinguish between the 20 amino acid residues, the index will be insufficient.

As shown in Figure 1a, the distribution values show a non-normal distribution, which is indicative of the non-gaussian process along the protein sequence. Roughly, the distribution looks like a U-shape where the highest probability of occurrence is obtained for the extreme values and the lowest for the mean value of the available D PRIFT index. Then, the pattern of the encoded protein sequence appears like complex bounced stairs with randomness as a sharp jump (Figure 1b).

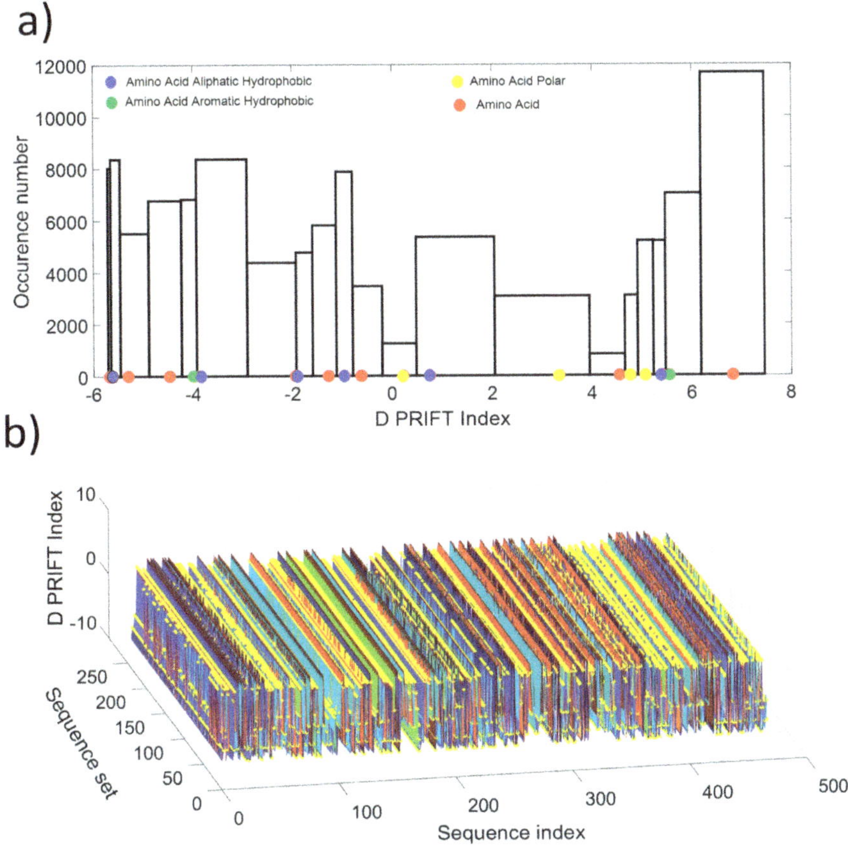

Figure 1. (**a**) Histogram of the D PRIFT index for 242 protein sequences. Red, blue, green, and yellow dots along the x-axis corresponds to the 20 values of the D PRIFT index. (**b**) Global view of the converted dataset (i.e., 242 protein sequences) using D PRIFT index rule. Yellow circle is indicative of the position within each sequence of the aliphatic hydrophobic, aromatic hydrophobic, and polar amino acids.

To target the jump stair pattern analysis within the protein sequence, we have used the normalized detrended cumulative sum (NDCS) method. The cumulative sum is a well-known and widely used algorithm to detect changes and shifts in time series [31]. In this study, we have extracted the linear long-term and normalized the cumulative sum of each sequence to (i) focus on the local change and (ii) have the same scale to compare transformed data. Figure 2 presents an example of transforming the original data (Sequence 1) into a detrended cumulative sum data. For clarity, we only present here the cumulative sum and linear detrending of the data. The normalized process is shown in the next figure. The trend of the cumulative sum is considered to be a linear trend for all the 242 protein sequences. The negative drift of the cumulative sum is related to the mean of a sequence. In our dataset, the average of the D PRIFT index is negative for each sequence and explains the downward drift of the cumulative sum.

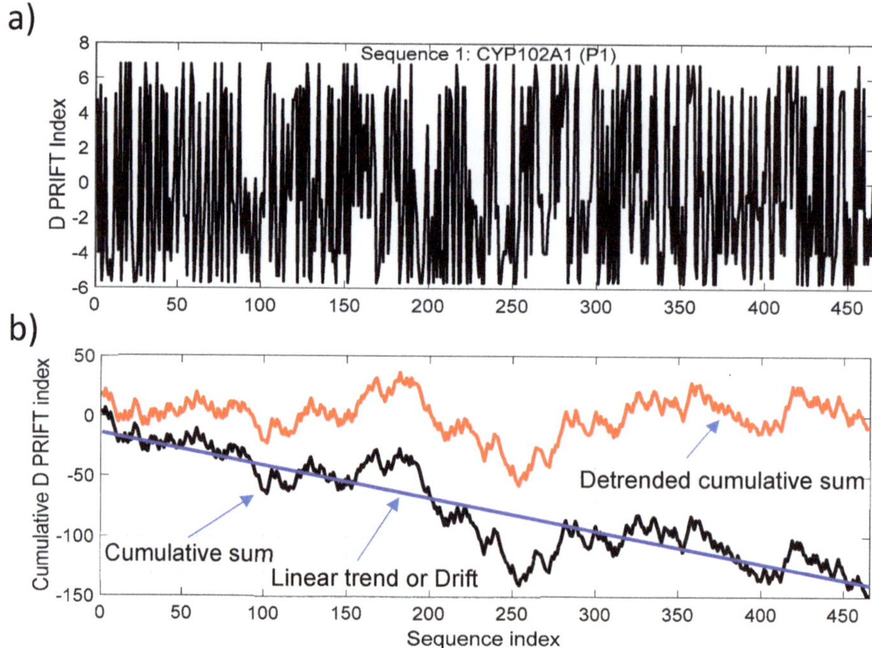

Figure 2. (**a**) D PRIFT index of sequence 1, which is parent CYP102A1 (P1); (**b**) Cumulative index (black line) and detrended cumulative sum (red line) of D PRIFT index of sequence 1. The blue line corresponds to the linear trend or drift of the cumulative sum of D PRIFT index.

Figure 3a depicts the NDSC plot in comparison with the original data (Sequence 1). Fluctuations reflect the local changes along the sequence and also a significant change pattern around the middle of the sequence. The fluctuation pattern relying on the cumulative sum transformation involves continuous distribution, conversely to the discrete distribution of the original D PRIFT index (Figure 3b).

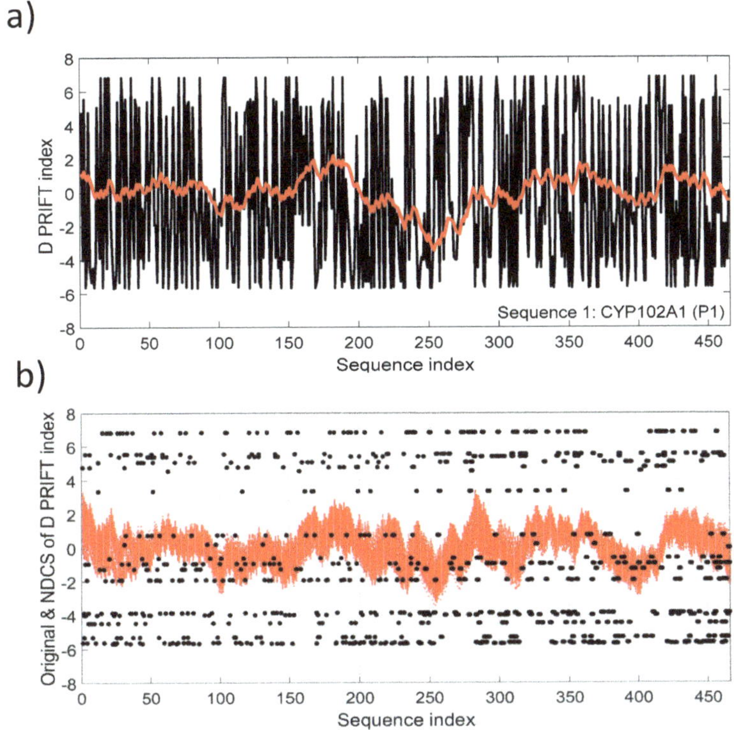

Figure 3. (**a**) D PRIFT index of sequence 1, which is parent CYP102A1 (P1). A superimposed red line corresponds to the normalized detrended cumulative sum (NDCS) of D PRIFT index; (**b**) Original (black dot) and normalized detrended cumulative sum (small red cross) of D PRIFT index for 242 protein sequences.

Figure 4a shows that the fluctuations of the NDCS of the D PRIFT index changes are normally distributed, with skewness close to 0 and kurtosis close to 3, which are the expected values for a normal distribution. In addition, the QQ-plot displayed in Figure 4b reveals that the observed distribution is close to a normal distribution and the two samples' (dataset values and generated normal data values) Kolmogorov–Smirnov test applied to this distribution does not reject the null hypothesis at the 5% significance level.

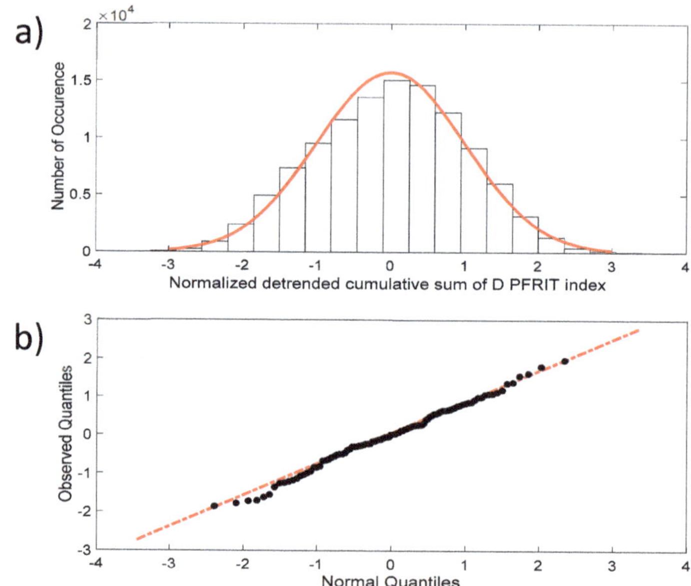

Figure 4. (**a**) Distribution of the NDCS of D PRIFT index changes for all sequences (black dots). Red line corresponds to Gaussian distribution; (**b**) QQ-plot of the NDCS of D PRIFT index changes quantiles and Gaussian quantiles. Red dotted line is a linear fitting of observed quantile distribution versus normal quantile distribution.

5. Results and Discussion

5.1. Normality and Intermittency

The changes along the protein sequence for four different pairwise distances show a platykurtic nature (Figure 5a). The average distribution exhibits large amplitude for fluctuations greater than 2.5 times the standard deviation of NDCS of D PRIFT index changes. The average is computed using 242 protein sequences. Below this threshold value, the distribution is close to the Gaussian distribution. This kind of departure from the Gaussian distribution in fluctuations is indicative of intermittency. Moreover, Figure 5b highlights that the platykurtic nature of the fluctuations covers a wide range of pairwise distances, but it is more pronounced with the [30–60] pairwise distance and for distances less than 10 pairwise. To summarize, this flat distribution indicates more diversity of changes for the large amplitude of pairwise distance within the protein sequence.

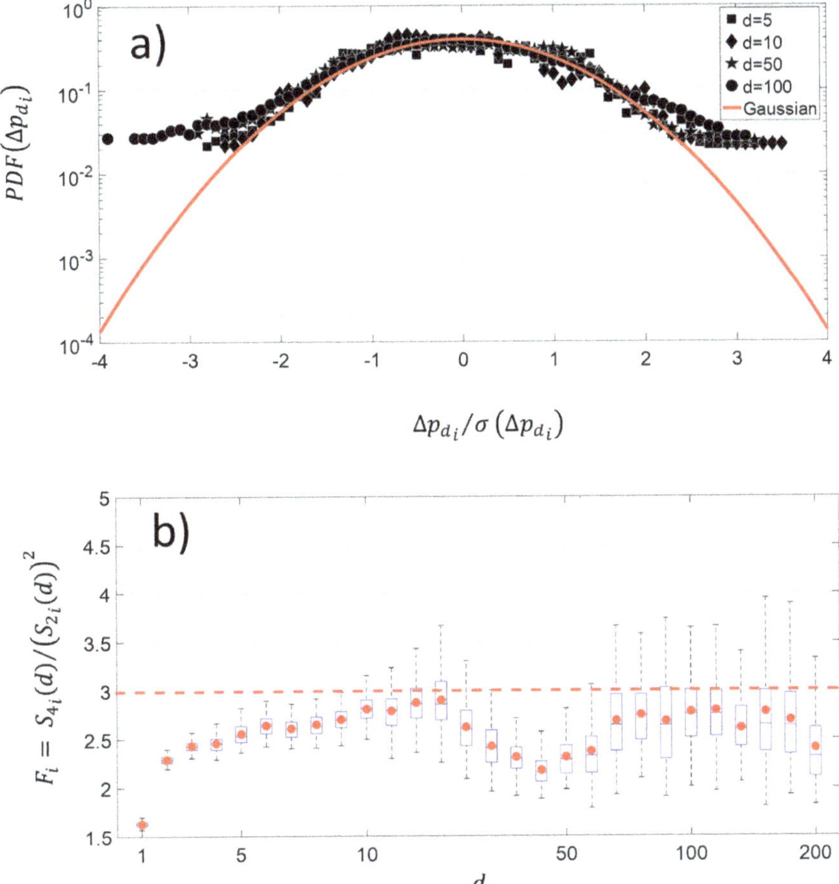

Figure 5. (**a**) Shape of average and normalized experimental probability functions (PDFs) of the increment of NDCS of PRIFT index changes at different distances in pairwise sequence $d = 5$, $d = 10$, $d = 50$, and $d = 100$ of 242 protein sequences. (**b**) Deviation of NDCS of PRIFT index changes distribution with respect to the Gaussian distribution at different pairwise sequence d.

5.2. Kolmogorov's Law and Brownian Process

We have conducted a Fourier analysis to focus on the fluctuation of the NDCS of D PRIFT index changes. Surprisingly, scale invariance can be detected in the log-log presentation of the Fourier spectra (Figure 6a). An average of -1.68 based on power law is obtained, which is very close to the Kolmogorov power law result of $-5/3$. This highlights that the fluctuations of the NDCS of D PRIFT index changes along a sequence are similar to a non-stationary process and obey the famous Kolmogorov's law of the energy cascade for turbulence in the inertial scale range [22]. In addition, as shown in Figure 6b, the range scale value for each sequence is rather close to $-5/3$, with an observed minimum slope value of -1.56 and a maximum slope value of -1.84. This means that the changes within the protein sequence can be formulated according to Fourier transform as $E(f) = f^{\beta}$ where β is the slope of the law and is close to the Kolmogorov spectrum. In addition, we can use criteria to check if the changes of protein are stationary or not [32]. This is summarized by the following test:

- $\beta < 1$, the changes are stationary,
- $\beta > 1$, the changes are non-stationary,

- $1 < \beta < 3$, the changes are non-stationary with stationary increments.

Thus, the changes in the sequence protein follow a non-stationary process. Moreover, the coefficient of variation of the fluctuations of the NDCS of D PRIFT index changes computed for all 242 sequences is less than 3%, confirming that this similarity with the Kolmogorov spectrum seems to be reproducible for each protein sequence as confirmed by the distribution of the spectrum slope obtained randomly with surrogated and shuffled data.

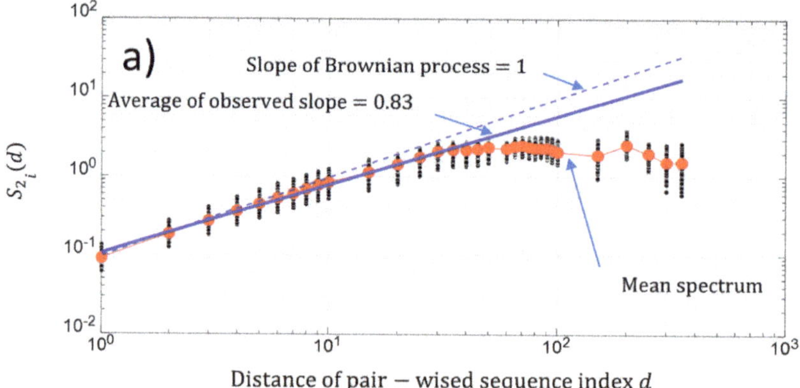

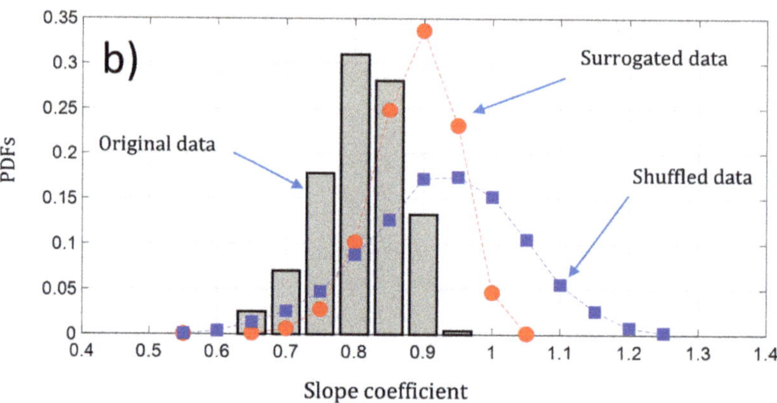

Figure 6. (**a**) Power spectrum density (PSD) of the NDCS of D PRIFT index changes of all 242 protein sequences S_i (black dots); (**b**) PDFs of the spectral exponent estimated from Fourier analysis. We have superimposed the PDFs obtained with surrogated (red spots) and shuffled data (blue squares).

As shown previously in Figure 3b, the fluctuations of the NDCS of D PRIFT index changes appear to show seemingly organized fluctuations. The question is *"Is there some dynamic pattern of these change fluctuations along a sequence S_i and is there some randomness of changes within the protein sequence?"*. A first approach is to analyze the behavior of the fluctuation of the pairwise protein index. Figure 7a shows that on average, the second-order moment $S_{2i}(d)$ of the pairwise protein sequence index separated by a distance d is linearly scaled in a sequence between pairwise protein sequence indexes separated by a distance d roughly below 50. We found a power law of 0.87, which is close to the Brownian power law process. Then, the behavior of the change fluctuations along each protein sequence S_i seems to be close to a Brownian process. Furthermore, we found for each protein sequence a power law between a range of [0.69 0.99] and a coefficient of variation less than 7%, which reveals that the fluctuations of NDCS of the D PRFIT index changes along a sequence S_i statistically have a behavior close to a Brownian process in regard to the results obtained with the surrogated and shuffled data (Figure 7b).

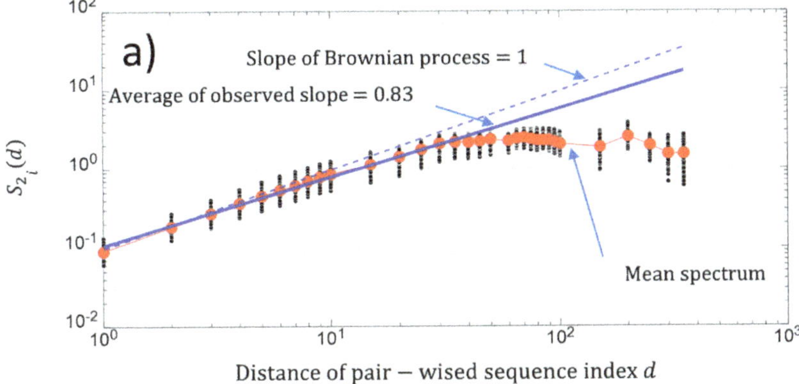

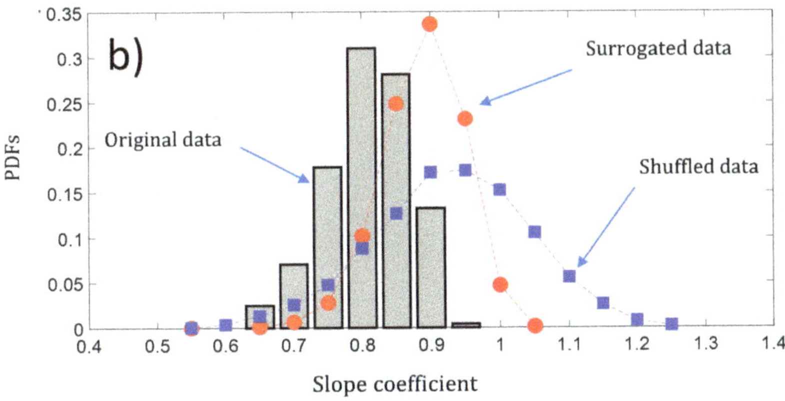

Figure 7. (a) Log-log presentation of the second-order moment $S_{2i}(d)$ of the NDCS of D PRIFT index changes of all 242 protein sequences S_{2i} versus the distance d of the pairwise protein sequence index (black dots); (b) PDFs of the slope of the scaling law distribution of the second-order moment $S_{2i}(d)$ of the NDCS of D PRIFT index changes estimated for each protein sequence S_i. We have superimposed the PDFs obtained with surrogated (red spots) and shuffled data (blue squares).

In addition, we have also computed the *q-order* moment for each protein sequence S_i. The result is shown in Figure 8a. As observed with second-order moment $S_{2i}(d)$ analysis, we again have a scaling law distribution between pairwise protein sequence index S_i below $d = 50$ for a higher-order moment. This result reveals the existence of a monofractal feature along the protein sequence S_i. Figure 8b shows that the fluctuations of NDCS of D PRIFT index changes of each protein sequence S_i contain a monofractal feature with $\xi(q) = 0.43\,q$, which is a linear law of q and reveals monofractal behavior. The slope of the linear law is called the Hurst exponent H. As a reminder, if the value of $H = \frac{1}{2}$, it means the changes in a sequence contain no memory as for the Brownian motion. If the changes of the sequence are anti-persistent $\left(0 < H < \frac{1}{2}\right)$, then the main pattern of the changes shows that a decrease is followed by an increase and vice-versa. Finally, if the Hurst exponent is as $\frac{1}{2} < H < 1$, then there is a persistent behavior in the changes and an increase or decrease will be maintained in a sequence. In our case, the changes are anti-persistent and they are statistically embedded between Kolmogorov process $\xi(q) = \frac{q}{3}$ [22] and the Brownian process $\xi(q) = \frac{q}{2}$. Thus, there is a potential stochastic model like the fractional Brownian model to predict the changes along the protein sequence.

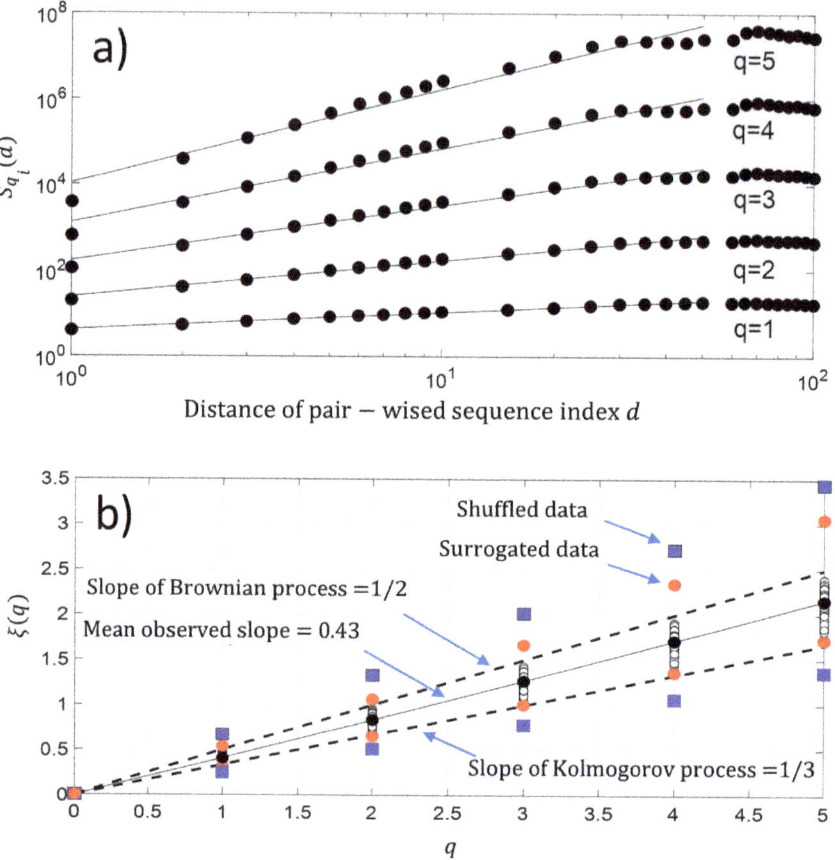

Figure 8. (**a**) Experimental high-order structure functions $S_{q_i}(d)$ with varying moments for $q = 1, 2, 3, 4$, and 5; (**b**) Generalized Hurst exponent $\xi(q)$. We have added the maximum and minimum value of $\xi(q)$ obtained with surrogated and shuffled data.

5.3. Entropy, Chaos, and Complexity

As previously mentioned, a sequence is defined as a set of alphabetic letters, which could be converted to other symbols (numerical, binary, etc.). Nevertheless, the changes of symbols or numerical values along the sequence are usually related to the real world of biochemical activities inside the whole protein sequence. The question is "*Do those changes present regular, irregular, chaotic and complex pattern within a sequence?*" Furthermore, nonlinear analysis is one approach to estimate the changes in features along a sequence. In this study, we have used five algorithms to assess the degree of the randomness or the disorder and complexity in protein sequences: (i) The Shannon entropy (*ShEn*); (ii) the sample entropy (*SampEn*); (iii) the largest Lyapunov exponent (*LLE*); (iv) Kolmogorov complexity (*KC*); and (v) the Kolmogorov complexity spectrum (*KCS*) algorithm. Table 2 presents the descriptive statistics of the NDCS of D PRIFT index changes for 242 protein sequences. On average, there is a significant amount of information in an entire probability distribution contained in a sequence. We observe that *SampEn* and *LLE* values are close to one. Moreover, the *KC* method underestimates the complexity in comparison to the *KCS* method, which takes into account the amplitude of the changes. Following the comparison with the surrogated and shuffled data generated from the original data, we found that the NDCS of D PRIFT index changes for 242 protein sequences used in this study include stochastic and moderate chaotic processes and show apparent embedding between the Kolmogorov ($H = 1/3$) and Brownian ($H = 1/2$) monofractal processes.

Table 2. Descriptive statistics of entropy, chaos, and complexity of the NDCS of D PRIFT index changes for 242 protein sequences.

D PRIFT Index		Entropy		Chaos	Complexity		Fractal
		Information	Regularity				
NDCS Data		Shannon Entropy	Sample Entropy	Largest Lyapunov Exponent	Kolmogorov Complexity	Kolmogorov Complexity Spectrum	Hurst Exponent
Minimum	Original	3.671	1.251	0.930	0.247	1.008	0.347
	Surrogate	3.514	1.051	0.730	0.152	1.046	0.332
	Shuffled	3.498	0.600	0.332	0.095	1.046	0.273
Mean	Original	3.880	1.433	1.277	0.475	1.071	0.432
	Surrogate	3.875	1.289	1.070	0.399	1.105	0.481
	Shuffled	3.911	1.147	0.911	0.328	1.103	0.498
Median	Original	3.888	1.436	1.286	0.475	1.065	0.436
	Surrogate	3.895	1.296	1.072	0.399	1.103	0.482
	Shuffled	3.933	1.154	0.906	0.323	1.103	0.498
Maximum	Original	4.066	1.618	1.601	0.647	1.141	0.481
	Surrogate	4.131	1.547	1.501	0.646	1.179	0.615
	Shuffled	4.188	1.604	1.469	0.627	1.160	0.690
Standard deviation	Original	0.084	0.063	0.117	0.084	0.031	0.027
	Surrogate	0.117	0.094	0.143	0.081	0.023	0.033
	Shuffled	0.130	0.188	0.220	0.109	0.022	0.058
1st quartile	Original	3.833	1.389	1.207	0.418	1.046	0.420
	Surrogate	3.805	1.226	0.969	0.342	1.084	0.459
	Shuffled	3.842	1.017	0.750	0.228	1.084	0.459
3rd quartile	Original	3.940	1.470	1.351	0.533	1.103	0.450
	Surrogate	3.963	1.355	1.160	0.456	1.122	0.503
	Shuffled	4.005	1.297	1.045	0.399	1.122	0.538

5.4. Drift (DRF), Kolmogorov Complexity Spectrum (KCS), and Activity (ACT): Linear Correlation and Superdiffusive Process between Sequences

The activity as defined in Section 2.2 (Thermostability) is also freely available for each protein sequence. Figure 9a shows the cumulative sum of activity, entropy, chaos, complexity, fractal, and drift parameters for 242 protein sequences. In order to track the biochemical activity changes through an invariant sequence arrangement, we have sorted, in ascending order, each sequence with increasing activity. Then, we have also sorted the remaining parameters in respect to the increasing activity and applied the cumulative sum. For clarity, we have presented the 10th of the entropy, chaos, complexity, fractal, and drift parameters, and the 1000th for activity. Most of the curves show a slightly linear shape, which is the average mode through increasing sequence activity. Nevertheless, the dynamic of changes through this increasing activity highlights that *NDCS*'s activity changes are well correlated with the *NDCS* of Kolmogorov complexity spectrum and drift (Figure 9b). There are pronounced parabola with an open upwards shape for activity (*ACT*) changes and a conversely open downwards shape for the Kolmogorov complexity spectrum (*KCS*) and drift (*DRF*) changes. The correlation coefficient is very high between *ACT*, *KCS*, and *DRF* as shown in Figure 9c.

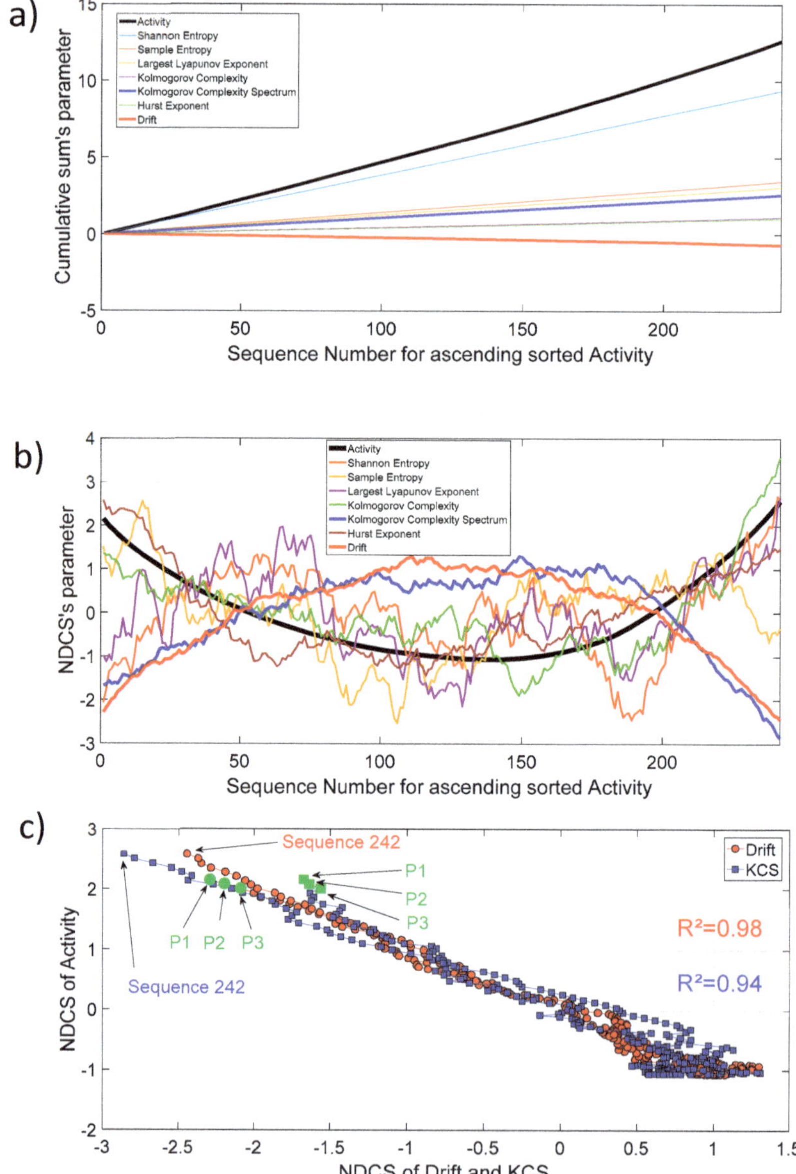

Figure 9. (**a**) Cumulative sum of activity, entropy, chaos, complexity, fractal, and drift parameters for ascending sorted activity; (**b**) Normalized detrended cumulative sum of activity, entropy, chaos, complexity, fractal, and drift parameters for ascending sorted activity; (**c**) NDCS of activity (*ACT*) versus NDCS of drift (*DRF*) and Kolmogorov complexity spectrum (*KCS*). The square of the correlation coefficient R^2 for both curves is added on the figure. The first and last sequence positions of the 242 ordered sequences are also shown. The green circle and square symbol indicate the position of the parents CYP102A1 (P1), CYP102A2 (P2), and CYP102A3 (P3) in this diagram.

We found a relationship between the inner-sequence changes drift, the complexity, and the activity throughout crossed 242 rearranged increasing activity protein sequences. As shown in Figure 9c, the trajectories of the bivariate parameter (drift, activity) or (complexity, activity) exhibits trajectories with jump between sequences, which leads to the question: *"Are these successive jumps related to variable changes ruled by a power law?"*. Then, we have analyzed these trajectories by calculating the mean square displacement of changes $\langle (\Delta d_S)^2 \rangle$ in the bivariate parameter (drift, activity) or (complexity, activity) space where d_S is the distance between two sequences. Moreover, we defined the mean square displacement as $\langle \Delta (d_S)^2 \rangle = \frac{1}{N_{dS}} \sum_{m=1}^{N_{dS}} \left[\left(X^j - X^k \right)^2 + \left(ACT^j - ACT^k \right)^2 \right]_{d_S = |k-j|}$ where N_{dS} is the number of pairwise sequences separated by a distance d_S and X is the drift (DRF) or Kolmogorov complexity spectrum (KCS). Figure 10 shows $\langle \Delta (d_S)^2 \rangle \sim d_S^{\alpha}$ with $\alpha \sim 1.7$ for the drift and $\alpha \sim 1.6$ for the complexity. We found that there is a scaling law of the bivariate (DFT, ACT) or (KCS, ACT) parameter that is similar to a super diffusive process with an exponent coefficient $\alpha > 1$ [33]. Here, we have plotted $\langle \Delta (d_S)^2 \rangle / \langle \Delta (d_{Sc})^2 \rangle$ where d_{Sc} is the characteristic distance between two sequences computed with the correlation function $\langle \delta (d_S) \rangle = \frac{1}{N_{dS}} \sum_{m=1}^{N_{dS}} \left[X^j X^k + ACT^j ACT^k \right]_{d_S = |k-j|}$.

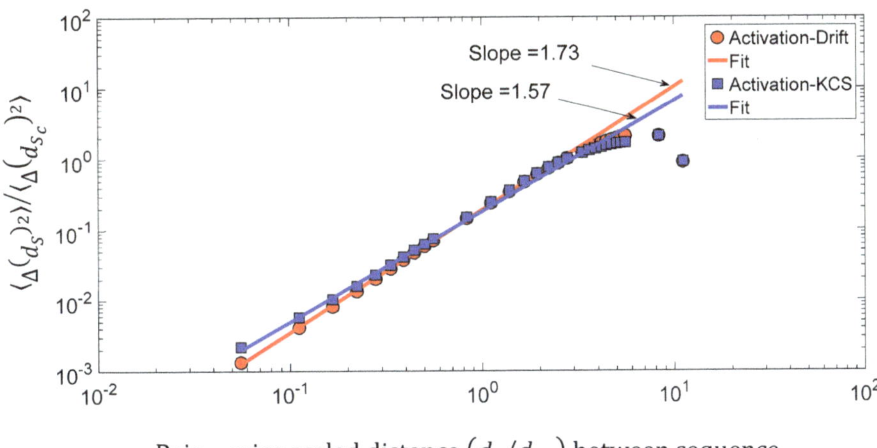

Pair − wise scaled distance $\left(d_S / d_{S_c} \right)$ between sequence

Figure 10. Log–log presentation of the mean square displacement $\langle \Delta (d_S)^2 \rangle / \langle \Delta (d_{Sc})^2 \rangle$ of the bivariate (KCS, ACT) parameter versus the pairwise scaled d distance (d_S / d_{Sc}) between sequences.

6. Conclusions

In this work, we analyze the nonlinear behavior of the D-PRIFT index changes around the overall linear trend scale of the protein sequence. To assess the nonlinear analysis, we have used protein residue values that are freely available, namely the AA index database. The protein dataset used contains 242 sequences and each sequence has 466 numerical values, one per amino acid residue. A protein sequence corresponds to a combination of encoding symbols from a dictionary of 20 standard amino acids symbols.

We have applied to each sequence a normalized detrended cumulative sum algorithm to extract the fluctuations of the numerical signal in the protein sequence. We analyzed these fluctuations with different tools, which are related to (i) entropy (information and regularity); (ii) chaos (largest Lyapunov exponent); (iii) complexity (Kolmogorov complexity and Kolmogorov complexity spectrum); and (iv) fractal (Hurst exponent). First, we showed that the change fluctuations of all the studied 242 protein sequences in the dataset seem to be non-stationary and follow on average a −5/3 Kolmogorov power-law. This result seems to be statistically significant in regard to a coefficient of variation less than 2% and a test done with randomly generated synthetically obtained data with surrogate and shuffle technique. To understand the nature of the inner changes within the protein sequence, we achieved the analysis of the variance of the changes through the scope of the spatial correlation: Here, the index position within the protein sequence. We found an invariance of pairwise scale index d, which is ruled by a $S_{2i}(d) \propto d^{\alpha}$ with $\alpha = 0.87$, a coefficient close to one of the well-known stochastic Brownian processes. The dispersion of the slope obtained for all 242 protein sequences is statistically coherent in comparison with the results obtained with synthetic data. Following the local analysis of the changes along the protein sequence, we have performed a systematic q-order moment of the fluctuations in order to track if there is a self-similar repeating pattern in the inner sequence. We showed that change fluctuations within the protein sequence have a monofractal behavior, which is an average among the 242 sequences embedded between the Kolmogorov and Brownian monofractal processes with a Hurst exponent ranging between $1/3$ and $1/2$. To encompass the local analysis and to have an overview of the nonlinearity analysis, we have computed statistical parameters related to entropy, chaos, complexity, and fractality. We demonstrated that the NDCS of D PRIFT index changes for the 242 protein sequences used in this study exhibit statistically moderate complexity, and low chaotic fluctuations.

Moreover, to integrate these results in the analysis of the protein activity changes for each sequence, we have conducted a study of the relationship between the linear-trend (drift) computed with the cumulative sum algorithm, the Kolmogorov complexity spectrum, which is indicative of computational complexity, and the activity of each protein sequence. As this analysis focused on the dynamics of the changes, we also applied the normalized detrended cumulative sum for these three parameters as done for the inner-sequence analysis. The results show a strong linear relationship between the bivariate (drift, activity) and (complexity, activity) parameters, which provides insight into the potential use of drift and complexity as a predictor in a linear model. Moreover, the analysis of the trajectories in the bivariate space highlights superdiffusive behavior of the change fluctuations with a power-law around -1.6 of the mean square displacement for each chosen bivariate parameter. This study demonstrates that the changes in the inner sequence and throughout the crossed inter-sequence are nonstationary, stochastic, irregular, complex, weakly chaotic, and monofractal. To conclude, there is some predictability of protein sequence changes, which can be modelled using a stochastic model. Linear law and scale invariance features found in this study should be explored in future work to study for classification, regression predictive model, and could be useful in the field of protein engineering.

Author Contributions: Data curation, X.F.C. and M.B.; formal analysis, M.B., X.F.C., S.P.C., and R.D.; investigation, X.F.C. and M.B.; methodology, M.B., and X.F.C.; project administration, M.B. and X.F.C.; resources, X.F.C. and M.B.; software, M.B. and X.F.C.; supervision, M.B.; validation, M.B., X.F.C., S.P.C., and R.D.; visualization, M.B. and X.F.C.; writing—original draft, X.F.C. and M.B.; writing—review and editing, M.B., X.F.C., S.P.C., and R.D.

Funding: Peaccel gratefully acknowledge support from a research program co-funded by the European Union (UE) and Region Reunion (FEDER). The funding agencies had no influence on the conduct of this research.

Conflicts of Interest: X.F.C. is linked to Peaccel. M.B., X.F.C., S.P.C. and R.D. declare no competing interests.

References

1. Hanson, J.; Yang, Y.; Paliwal, K.; Zhou, Y. Improving protein disorder prediction by deep bidirectional long short-term memory recurrent neural networks. *Bioinformatics* **2016**. [CrossRef] [PubMed]
2. Kovacs, A.; Telegdy, G. Modulation of active avoidance behavior of rats by ICV administration of CGRP antiserum. *Peptides* **1994**, *15*, 893–895. [CrossRef]
3. Niessen, K.A.; Xu, M.; George, D.K.; Chen, M.C.; Ferré-D'Amaré, A.R.; Snell, E.H.; Cody, V.; Pace, J.; Schmidt, M.; Markelz, A.G. Protein and RNA dynamical fingerprinting. *Nat. Commun.* **2019**, *10*, 1026. [CrossRef] [PubMed]
4. Qi, Z.-H.; Jin, M.-Z.; Li, S.-L.; Feng, J. A protein mapping method based on physicochemical properties and dimension reduction. *Comput. Biol. Med.* **2015**, *57*, 1–7. [CrossRef] [PubMed]
5. Gök, M.; Koçal, O.H.; Genç, S. Prediction of Disordered Regions in Proteins Using Physicochemical Properties of Amino Acids. *Int. J. Pept. Res. Ther.* **2016**, *22*, 31–36. [CrossRef]

6.   Wang, Y.; You, Z.H.; Yang, S.; Li, X.; Jiang, T.H.; Xi, Z.X. A High Efficient Biological Language Model for Predicting Protein–Protein Interactions. *Cells* **2019**, *8*, 122. [CrossRef] [PubMed]

7.   Plötz, T.; Fink, G.A. Pattern recognition methods for advanced stochastic protein sequence analysis using HMMs. *Pattern Recognit.* **2006**, *39*, 2267–2280. [CrossRef]

8.   Chattopadhyay, A.K.; Nasiev, D.; Flower, D.R. A statistical physics perspective on alignment-independent protein sequence comparison. *Bioinformatics* **2015**, *31*, 2469–2474. [CrossRef] [PubMed]

9.   Vinga, S. Information theory applications for biological sequence analysis. *Brief. Bioinform.* **2014**, *15*, 376–389. [CrossRef]

10.   Zhao, J.; Wang, J.; Hua, W.; Ouyang, P. Algorithm, applications and evaluation for protein comparison by Ramanujan Fourier transform. *Mol. Cell. Probes* **2015**, *29*, 396–407. [CrossRef]

11.   Czerniecka, A.; Bielińska-Wąż, D.; Wąż, P.; Clark, T. 20D-dynamic representation of protein sequences. *Genomics* **2016**, *107*, 16–23. [CrossRef] [PubMed]

12.   Zhu, X.-J.; Feng, C.-Q.; Lai, H.-Y.; Chen, W.; Hao, L. Predicting protein structural classes for low-similarity sequences by evaluating different features. *Knowl. Based Syst.* **2019**, *163*, 787–793. [CrossRef]

13.   Yang, L.; Wei, P.; Zhong, C.; Meng, Z.; Wang, P.; Tang, Y.Y. A Fractal Dimension and Empirical Mode Decomposition-Based Method for Protein Sequence Analysis. *Int. J. Pattern Recognit. Artif. Intell.* **2019**. [CrossRef]

14.   Yu, J.F.; Cao, Z.; Yang, Y.; Wang, C.L.; Su, Z.D.; Zhao, Y.W.; Wang, J.H.; Zhou, Y. Natural protein sequences are more intrinsically disordered than random sequences. *Cell. Mol. Life Sci.* **2016**, *73*, 2949–2957. [CrossRef] [PubMed]

15.   Cao, C.; Liu, F.; Tan, H.; Song, D.; Shu, W.; Li, W.; Zhou, Y.; Bo, X.; Xie, Z. Deep Learning and Its Applications in Biomedicine. *Genom. Proteom. Bioinform.* **2018**, *16*, 17–32. [CrossRef]

16.   Li, Y.; Drummond, D.A.; Sawayama, A.M.; Snow, C.D.; Bloom, J.D.; Arnold, F.H. A diverse family of thermostable cytochrome P450s created by recombination of stabilizing fragments. *Nat. Biotechnol.* **2007**, *25*, 1051–1056. [CrossRef] [PubMed]

17.   Kawashima, S.; Ogata, H.; Kanehisa, M. Aaindex: Amino Acid Index Database. *Nucleic Acids Res.* **1999**, *27*, 368–369. [CrossRef]

18.   Kawashima, S.; Pokarowski, P.; Pokarowska, M.; Kolinski, A.; Katayama, T.; Kanehisa, M. Aaindex: Amino acid index database, progress report 2008. *Nucleic Acids Res.* **2008**, *36*, D202–D205. [CrossRef]

19.   Shannon, C.E. A Mathematical theory of Communication. *Bell Syst. Tech. J.* **1948**, *27*, 379–423. [CrossRef]

20.   Richman, J.S.; Moorman, J.R. Physiological time-series analysis using approximate entropy and sample entropy. *Am. J. Physiol. Heart Circ. Physiol.* **2000**, *278*, H2039–H2049. [CrossRef]

21.   Wolf, A.; Swift, J.B.; Swinney, H.L.; Vastano, J.A. Determining Lyapunov exponents from a time series. *Phys. Nonlinear Phenom.* **1985**, *16*, 285–317. [CrossRef]

22.   Kolmogorov, A.N. The local structure of turbulence in incompressible fluid for very large Reynolds numbers. *Dokl. Akad. Nauk. SSSR* **1941**, *30*, 299–303. [CrossRef]

23.   Chaitin, G.J. On the Length of Programs for Computing Finite Binary Sequences: Statistical considerations. *J. ACM* **1969**, *16*, 145–159. [CrossRef]

24.   Lempel, A.; Ziv, J. On the Complexity of Finite Sequences. *IEEE Trans. Inf. Theory* **1976**, *22*, 75–81. [CrossRef]

25.   Mihailović, D.T.; Mimić, G.; Nikolić-Djorić, E.; Arsenić, I. Novel measures based on the Kolmogorov complexity for use in complex system behavior studies and time series analysis. *Open Phys.* **2015**, *13*, 1–14. [CrossRef]

26.   Ziv, J.; Lempel, A. Compression of individual sequences via variable-rate coding. *IEEE Trans. Inf. Theory* **1978**, *24*, 530–536. [CrossRef]

27.   Monin, A.S.; Yaglom, A.M. *Statistical Fluid Mechanics: Mechanics of Turbulence*; MIT Press: Cambridge, MA, USA, 1987; Volume 1, p. 784.

28.   Schreiber, T.; Schmitz, A. Surrogate time series. *Phys. Nonlinear Phenom.* **2000**, *142*, 346–382. [CrossRef]

29.   Peng, C.; Buldyrev, S.V.; Havlin, S.; Simons, M.; Stanley, H.E.; Goldberger, A.L. Mosaic organization of DNA nucleotides. *Phys. Rev. E* **1994**, *49*, 1685–1689. [CrossRef] [PubMed]

30.   Cornette, J.L.; Cease, K.B.; Margalit, H.; Spouge, J.L.; Berzofsky, J.A.; DeLisi, C. Hydrophobicity scales and computational techniques for detecting amphipathic structures in proteins. *J. Mol. Biol.* **1987**, *195*, 659–685. [CrossRef]

31. Regier, P.; Briceño, H.; Boyer, J.N. Analyzing and comparing complex environmental time series using a cumulative sums approach. *MethodsX* **2019**, *6*, 779–787. [CrossRef]
32. Marshak, A.; Davis, A.; Cahalan, R.; Wiscombe, W. Bounded cascade models as nonstationary multifractals. *Phys. Rev. E* **1994**, *49*, 55–69. [CrossRef] [PubMed]
33. Richardson, L.F. Atmospheric Diffusion Shown on a Distance-Neighbour Graph. *Proc. R. Soc. Math. Phys. Eng. Sci.* **1926**, *110*, 709–737. [CrossRef]

article

ELIHKSIR Web Server: Evolutionary Links Inferred for Histidine Kinase Sensors Interacting with Response Regulators

Claude Sinner [1,†], Cheyenne Ziegler [1,†], Yun Ho Jung [1], Xianli Jiang [1] and Faruck Morcos [1,2,3,*]

1. Department of Biological Sciences, University of Texas at Dallas, Richardson, TX 75080, USA; claude.sinner@utdallas.edu (C.S.); cheyenne.ziegler@utdallas.edu (C.Z.); yxj180001@utdallas.edu (Y.H.J.); xianli.jiang@utdallas.edu (X.J.)
2. Center for Systems Biology, University of Texas at Dallas, Richardson, TX 75080, USA
3. Department of Bioengineering, University of Texas at Dallas, Richardson, TX 75080, USA
* Correspondence: faruckm@utdallas.edu
† These authors contributed equally to this work.

Abstract: Two-component systems (TCS) are signaling machinery that consist of a histidine kinases (HK) and response regulator (RR). When an environmental change is detected, the HK phosphorylates its cognate response regulator (RR). While cognate interactions were considered orthogonal, experimental evidence shows the prevalence of crosstalk interactions between non-cognate HK–RR pairs. Currently, crosstalk interactions have been demonstrated for TCS proteins in a limited number of organisms. By providing specificity predictions across entire TCS networks for a large variety of organisms, the ELIHKSIR web server assists users in identifying interactions for TCS proteins and their mutants. To generate specificity scores, a global probabilistic model was used to identify interfacial couplings and local fields from sequence information. These couplings and local fields were then used to construct Hamiltonian scores for positions with encoded specificity, resulting in the specificity score. These methods were applied to 6676 organisms available on the ELIHKSIR web server. Due to the ability to mutate proteins and display the resulting network changes, there are nearly endless combinations of TCS networks to analyze using ELIHKSIR. The functionality of ELIHKSIR allows users to perform a variety of TCS network analyses and visualizations to support TCS research efforts.

Keywords: statistical inference; mutational phenotypes; interaction specificity; phosphorylation; fitness landscape; bacterial signaling

Citation: Sinner, C.; Ziegler, C.; Jung, Y.H.; Jiang, X.; Morcos, F. ELIHKSIR Web Server: Evolutionary Links Inferred for Histidine Kinase Sensors Interacting with Response Regulators. *Entropy* **2021**, *23*, 170. https://doi.org/10.3390/e23020170

Academic Editor: Alessandro Giuliani

Received: 17 December 2020
Accepted: 26 January 2021
Published: 30 January 2021

Publisher's Note: MDPI stays neutral with regard to jurisdictional claims in published maps and institutional affiliations.

1. Introduction

Two-component systems (TCSs) are ubiquitous in bacteria and archaea and are the key signaling transduction machineries for sensing and responding to the environment. TCSs consist of sets of interaction signaling partners, histidine kinases (HKs) that phosphorylate their cognate response regulators (RRs). Interactions, however, are often not one-to-one. Multiple HKs can interact with multiple RRs. Identifying relevant interactions among TCS is an important task that has been addressed experimentally only for a limited number of organisms.

We advanced the study of interaction specificity in TCS by creating a model based on amino acid coevolution at the interface of HKs and RRs. Our Direct Coupling Analysis (DCA) [1] based interaction model not only confirms known cognate partners [2] but also reveals novel interactions in multiple organisms. We uncovered a TCS network in Synechococcus elongatus regulating cyanobacterial circadian clock and confirmed important master regulators [3]. Our model is also able to predict functional mutations to modulate binding specificity between partners, such as PhoQ and PhoP [4] or even design new interactions between non-cognate, interspecies TCS proteins, such as the EnvZ from Escherichia coli and Spo0F from Bacillus subtilis [5]. Another application of this model is the identification of crosstalk across signaling networks and the influence of mutation in the topology of the network. Figure 1 illustrates a section of statistical couplings in a protein sequence and highlights two of the most common applications, the identification of physical contacts in a protein [6,7] or the identification and quantification of interaction between multiple proteins [8,9].

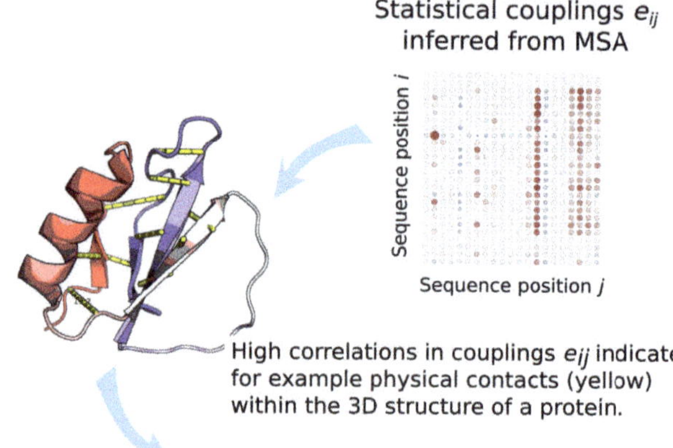

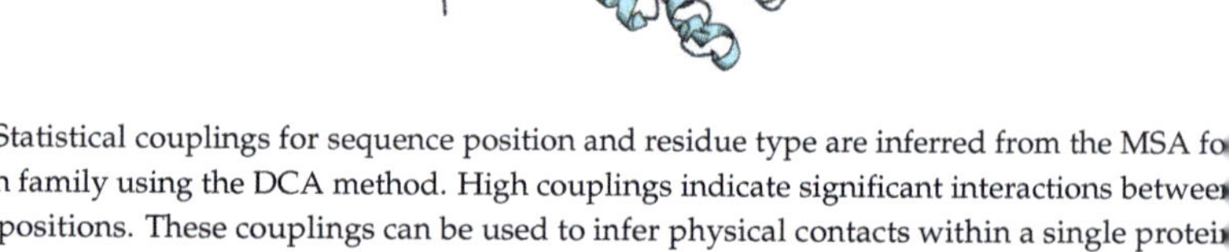

Figure 1. Statistical couplings for sequence position and residue type are inferred from the MSA for the protein family using the DCA method. High couplings indicate significant interactions between sequence positions. These couplings can be used to infer physical contacts within a single protein structure, or to infer the interaction interface and strength between two proteins.

We decided to make this model and tools available to the scientific community in an interactive web server that facilitates the analysis and prediction of TCS networks as well as the exploration of the effects of mutation in these proteins prior to experimental work. We named the service Evolutionary Links Inferred for Histidine Kinase Sensors Interacting with Response regulators (ELIHKSIR) and it can be accessed at https://elihksir.org.

In recent years, online repositories of sequence data have seen a large influx of sequences and are painting a more refined picture of protein families. Using these data one can construct global probabilistic models that verify the observed statistics and relate them to inter-residual couplings. Cheng et al. [2] have used these probabilistic models to introduce an objective function $H_{TCS}^{specific}(\vec{\sigma})$ to describe the specificity (fitness) of the interaction between a response regulator and a histidine kinase partner by a scalar score using a sequence $\vec{\sigma}$ from a linked multiple sequence alignment (MSA). For completeness we reproduce the introduction of $H_{TCS}^{specific}(\vec{\sigma})$ here.

Using the set of sequences $\{\vec{\sigma}\}$, we can create a global probabilistic model $P(\vec{\sigma})$ to find a given amino acid sequence $\vec{\sigma}$ in a protein family by the following:

$$P(\vec{\sigma}) = \frac{1}{Z} \cdot \exp\left(-H(\vec{\sigma})\right) \tag{1}$$

with a general Hamiltonian $H(\vec{\sigma})$ and the partition function Z to verify normalization for the probabilities. A sufficient form for $H(\vec{\sigma})$ [10] is given by the large-q Potts Model [11]:

$$H(\vec{\sigma}) \propto -\sum_{ij} e_{ij}(a_i, a_j) - \sum_{i} h_i(a_i) \tag{2}$$

with the coupling matrix $e_{ij}(a_i, a_j)$ between two sequence sites a_i, a_j at sequence positions i and j; and the local field $h_i(a_i)$ at the site a_i at sequence position i. The sites a can have $q = 21$ different states for amino acid and sequence gap composition. The entries of the coupling matrix $e_{ij}(a_i, a_j)$ and the local fields $h_i(a_i)$ encode preferences for sequence compositions at positions i and j. The inference of the coupling matrix $e_{ij}(a_i, a_j)$ and the local fields $h_i(a_i)$ is a non-trivial task. Several methods exist to do so [1,12,13]. We inferred the couplings using mean field DCA (mfDCA), which is fast and accurate at predicting interaction specificity in TCS.

From these coupling parameters, we can introduce and create objective functions to measure varying effects. In the Material and Methods, we introduce an objective function $H_{TCS}^{specific}(\vec{\sigma})$ that is sensitive to sequence mutations and linked to protein interaction specificity. For the calculation of $H_{TCS}^{specific}(\vec{\sigma})$, we need full access to the couplings $e_{ij}(a_i, a_j)$ and local fields $h_i(a_i)$. Throughout the process, we consider these as constant and created a database that our server uses internally to calculate new values for the $H_{TCS}^{specific}(\vec{\sigma})$ score in a mutation event.

Figure 2 gives an overview of the entire process of the ELIHKSIR web server. The MSA for our system is created by concatenating the HisKA domain section of the Pfam [14] Histidine Kinase (HK) family (Pfam:PF00512) [15] and the REC domain of the Response Regulator (RR) family (PF00072) [16], which contains information for thousands of organisms. Furthermore, we collect metadata for each organism and sequence pairs through the Uniprot database [17]. From this, we calculate the coupling matrices $e_{ij}(a_i, a_j)$ and the local fields $h_i(a_i)$. These parameters allow us to calculate a score for the interaction specificity H_{TCS}. The data are visualized in a web interface with interactive heatmaps.

ELIHKSIR is a user-friendly and accessible tool that displays TCS signaling networks. The breadth of the web server allows for analysis of TCS networks in both common and uncommon species and strains. Table 1 summarizes the number of organisms and interaction partners available. Users can easily search for their organism of interest, view TCS specificity networks for the whole organism, and view all possible interactions for an HK or RR of interest. This capability allows researchers with restricted computational resources to analyze signaling networks. Some common use cases of ELIHKSIR's features include identifying cross-talk interactions between non-cognate HKs and RRs, comparing specificity of different HK–RR pairs, and comparing differences in signaling networks between species and/or strains. In addition to browsing and exporting wild-type TCS networks, mutations may be introduced into HKs and/or RRs, for which all interaction specificity scores are recalculated and displayed. This allows users to predict network-wide changes in specificity after introducing a mutation. Further applications include testing mutants for desired change(s) in specificity, guiding engineering of TCS proteins with interaction or insulation requirements, and viewing changes in specificity for new or uncommon clinical and environmental variants. With these capabilities, ELIHKSIR is an effective tool for a variety of researchers who interface with TCS proteins and signaling.

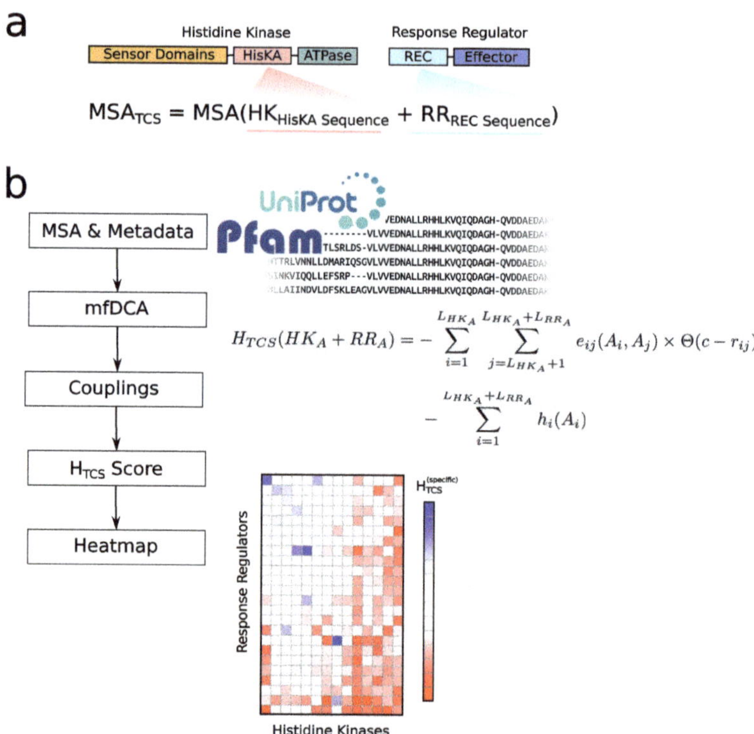

Figure 2. (**a**) A concatenated MSA is generated for Pfam [14] Histidine Kinase (HK) fami (Pfam:PF00512) [15] and Response Regulator (RR) family (PF00072) [16]. (**b**) From this MSA coupli matrices are generated with mfDCA [1]. From these couplings, we are able to calculate a numer score using the equation shown. This equation formally describes how Hamiltonian scores a generated for each HK–RR pair and is equivalent to Equation (3). The data are displayed in a we interface with interactive heatmaps. The user has an elaborate menu available to explore the data creating mutations to sequence positions. The default heatmap legend is more sensitive towards t outer extremes of the values, coloring strongly negative (favorable) or positive values (unfavorabl

Table 1. Attributes of the ELIHKSIR web server.

Total Organisms	66
Bacteria	64
Archaea	6
Eukaryotes	1
Unknown Organisms/Metagenomes	1
Total Interactions Evaluated	6,272,6
Number of HKs	111,0
Number of RRs	225,6

2. Results

2.1. Validation

Validation of the ELIHKSIR web server was performed through detailed investigatic using three model organisms: Escherichia coli, Synechococcus elongatus, and Enterococcu faecalis. True positive specificity predictions were defined by either positive selectic and/or negative selection for a cognate pair. Positive selection is defined as an HK havir its highest specificity with a single RR. Negative selection is defined as an RR having poor specificity across all HKs but having its relative highest specificity with an HK. False negatives were defined as selection towards a noncognate partner that is greater than that of the cognate partner, in which both positive and negative selection fail to identify the cognate pair. Only cognate pairs in which the HK contains a HisKA domain were evaluated. For *E. coli*, there were fourteen true positives and three false negatives for seventeen cognate pairs, shown in Figure A1. For *S. elongatus*, there were five true positives and one false negative for the six cognate pairs, shown in Figure A2. For *E. faecalis*, there were seven true positives and one false negative for the eight cognate pairs, shown in Figure A3. The resulting sensitivity and accuracy is 0.84.

DCA identifies coevolving residues at the HK–RR interface for HisKA and REC domains that have been used to accurately predict the structure of the HK–RR complex [18]. Out of the top 20 DCA-identified interfacial couplings, 10 are present in the 3DGE structure, as shown in Figure A4b. Information about all 3DGE interfacial contacts is present in the DCA-generated couplings and local fields (Figure A4a). Couplings are scored by their direct information (DI) value as defined by DCA (Table A2). Thus, higher DI values indicate that these couplings are more important for HK–RR interactions. When utilizing DCA couplings for the calculation of Hamiltonian values, only couplings present on the structurally verified HK–RR interface are used. This ensures auxiliary information obtained through DCA does not impact the Hamiltonian values, and thus, does not impact the resulting specificity score.

The interface is aligned for each TCS pair during the construction of the MSA, which was performed using a hidden Markov model. The sequences displayed in ELIHK-SIR are the aligned residues and gaps. Predictions made based on HK and RR sequences only consider residues which align with their respective protein family. Insertions and deletions are not considered in the alignment of the interface and may result in deviations in the three dimensional structure of the resulting signaling complex. The model assumes no changes in the three dimensional structure of the HK–RR interface during evaluation of different TCS pairs.

2.2. Mutations

A key functionality of the ELIHKSIR server is the ability to interactively perform in silico mutations on a HK–RR pair. In the mutation screen, as shown in Figure 3b, the full MSA of a pair is shown with a visual clue to the histidine kinase region and the response regulator region. Any part of the MSA can be transformed and the changes in a HK or RR become applied globally. The heatmap is also updated accordingly. Gaps can be introduced as '-' characters. As the mutation values are run against a tabulated database for the positions and amino acid type, the total length of the MSA has to remain at 176 amino acids. Insertions are not possible in the model unless they occur in gap regions.

Only a subset of the positions in the genetic sequence correspond to an actual interfacial residue of the protein interface between Thermotoga maritima class I HK853 and the response regulator RR468, (PDB ID: 3DGE). Because of this, not every change in the sequence performed by a user will translate into a change in the specificity score. Furthermore, some types of amino acids can play similar roles in a specific residue position. In this case, the model accounts for this and only reflects minor or no changes in the total score.

An interesting application of the mutation user interface is shown in Figure 4, the rewiring of specificity. By transferring portions of a sequence from one cognate pair to another cognate pair, interaction properties can be discovered or lost. In this specific example, a portion of amino acids positions 70 to 80 transferred from ntrC into the same position in the cusR response regulator creates cross-talk with a new interacting partner qseC, while maintaining the initial interaction cognate partner cusS. Alternatively, introducing the same sequence positions from the response regulator qseB into cusR is entirely sufficient to rewire the entire interaction and create an exclusively positive selection towards qseC.

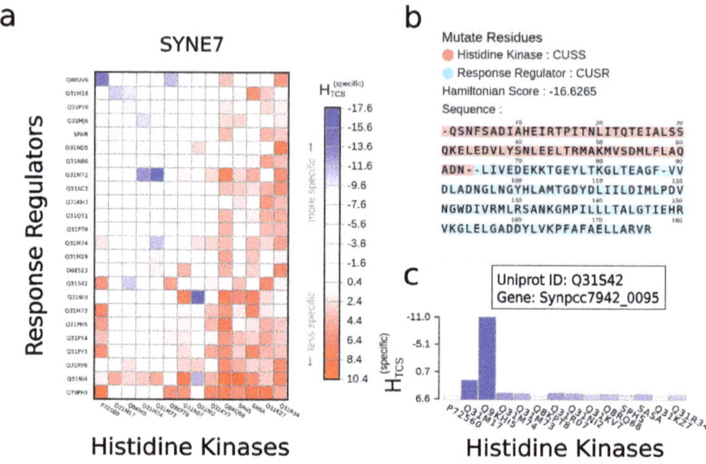

Syne7: Synechococcus elongatus (strain PCC 7942)

Figure 3. (**a**) Heatmap for Synechoccus elongatus as displayed on ELIHKSIR and when exported as an image. (**b**) Mutation screen as displayed on ELIHKSIR. (**c**) Histogram depicting all selectivity scores for a given HK or RR.

ECOLI: (Escherichia coli)

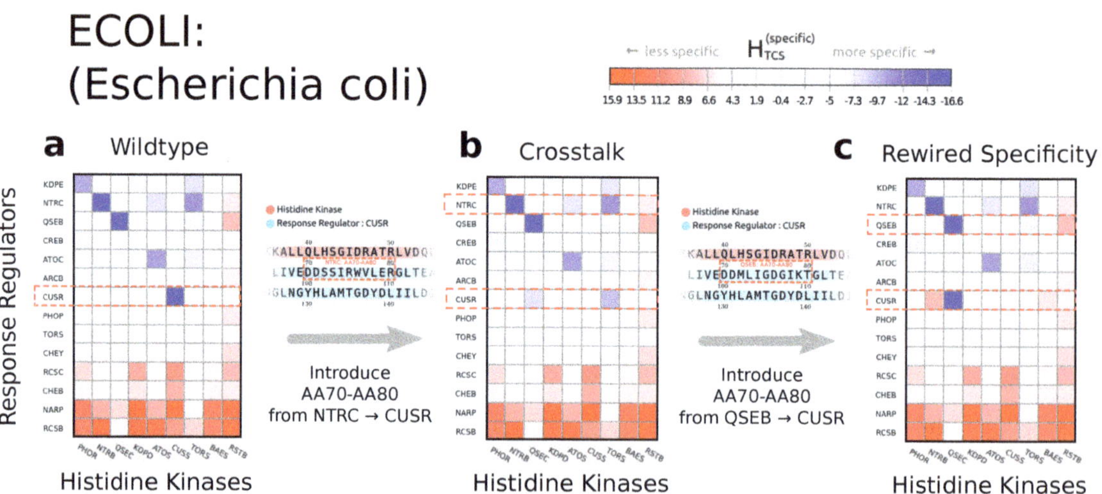

Figure 4. (**a**) One of the many use cases for the web server is the exploration and in silico change of specificity. In this example, we identify the response regulator cusR as the interaction partner of the histidine kinase cusS indicated by the lowest value in our Hamiltonian. (**b**) The transfer of a significant sequence portion of the response regulator ntrC does not disrupt the initial interaction and introduces cross-talk through a second interaction partner. (**c**) Alternatively, the introduction of a sequence portion of the response regulator qseB into cusR disrupts the initial interaction and rewires the interaction towards qseC.

2.3. Data Export

The user has three options to export data from ELIHKSIR. First, the user may export a PNG image, as shown in Figure 3a of the entirety of the heatmap in PNG format by clicking on the Export to PNG button on the left panel once a heatmap has been displayed. This will generate a PNG image of the heatmap on a transparent background and download it onto the user's machine. The image will also include the labels and legend. When selecting an $n \times m$-sized subselection in a heatmap, the user is presented with the choice to display the subselection as a new heatmap. Second, the user may export a PNG image of a histogram as shown in Figure 3c of a row of response regulator and histidine kinase pairs that corresponds to a desired histidine kinase by clicking on the *Export to PNG* button that is located inside the opened histogram. The histogram export will also include the names of each response regulator. Finally, the user may export a CSV representation of the user's arbitrary selections of the cells of the heatmap. After the user makes selections of the cells on the heatmap, the *Export to CSV* button on the right panel can be clicked to download a file that contains a comma delimited list of the user's selections. All these methods of exporting will take into consideration the mutated Hamiltonian values, if any, of the response regulator and histidine kinase pairs.

2.4. Negative Selection

An important concept highlighted by the server is that of negative selection. Not only are interaction partners indicated by strong couplings and a highly negative score for a TCS pair, but equally by high interaction scores with each partner except one. In this case, the interaction with a marginal advantage will be the strongest interaction and may facilitate signal transduction. Hence, we differentiate by either positive selection and/or negative selection for the cognate pair, where positive selection is defined as an HK having the highest specificity for its cognate RR and where negative selection is defined as the cognate HK having the highest specificity out of all HKs for a given RR. Figure 5 highlights this for two different cases in *E. coli* (ECOLI). Besides the heatmap, a good indicator for the interactions is a look at the histograms (Figure 5b) of interaction strengths, which are, for this purpose, available through the server. In cusR, a single interaction between cusR and the histidine kinase cusS is dominant (Figure 5b top). In rcsB, the majority of interactions are reported as less specific. Even though the interaction between rcsB and the histidine kinase rcsC is not reported as very specific, it will be the dominant interaction for rcsB.

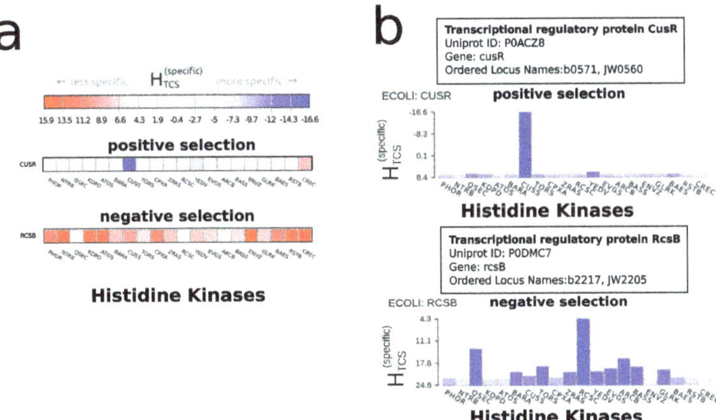

Figure 5. Negative selection in Escherichia coli strain K12 (ECOLI). (**a**) Heatmap view for the response regulators cusR and rcsB. In cusR, a single interaction between cusR and the histidine kinase cusS is dominant. This is a case of positive selection between two interacting partners. In rcsB, the majority of interactions are reported as having a low specificity. Even though the interaction between rcsB and the histidine kinase rcsC is not reported as having a high specificity, it will be the dominant interaction for rcsB as there is no stronger interaction partner for signal transduction. This is an example of negative selection. (**b**) Histogram view for the response regulators cusR and rcsB. From these histograms, it becomes clear that cusR-cusS (top) and rcsB-rcsC (bottom) are the dominant interactions.

3. Discussion

3.1. Characterization of Cognate Specificity

Through both mutational and computational analyses, the interface between the HisKA domain and the REC response regulator domain has been shown to control specificity of TCS interactions [19]. In Figure 6, this finding is confirmed for 14 out of 17 cognate pairs shown for *E. coli*. In Figure 7, this finding is confirmed for all eight cognate pairs shown for M. tuberculosis. While predictions of interaction specificity have been previously demonstrated, ELIHKSIR presents specificity scores for all HisKA HK and RR pairs in thousands of organisms, defining specificity landscapes. These specificity landscapes can then be used to determine favorable interactions through identification of pairs exhibiting positive and/or negative selection. When assessing cognate pairs, the prevalence of interactions either partially or solely characterized by negative selection becomes apparent. In the validation process, 54.8% of detected cognate pairs exhibited both positive and negative selection and 19.4% of detected cognate pairs were characterized by negative selection only. Negative selection is important for preventing cross-talk and ensuring orthogonality [20] but results indicate that it may be a main or contributing determinant of many cognate interactions. It is unclear if other attributes or domains contribute to reinforcement of specificity for cognate pairs detected by negative selection only.

By identifying whether cognate interactions are maintained by positive and/or negative selection, users can explore how deletion of TCS proteins may affect gene expression. Experimental deletion of the cognate RR in a pair regulated by negative selection may result in a noncognate RR being phosphorylated by the HK. In deletion experiments, it may be useful to understand how removal of TCS proteins may affect overall expression. Furthermore, some TCS proteins are encoded for on plasmids. Understanding how the presence or lack of plasmid-encoded TCS proteins on organisms' genetic expression may be important for the study of antibiotic resistance and plant cell transformation by bacteria [21].

It is important to note that, in many proteins, HisKA domains are accompanied by an HATPase_c domain, which is responsible for binding ATP and transferring its -phosphate to the HisKA domain. Aside from its ATPase activity, the HATPase_c domain alone can act as a histidine kinase [22]. It is unknown whether the HATPase_c domain itself encodes specificity or is partially responsible for specificity in certain cognate TCS pairings. Further analysis of the HATPase_c domain as well as other histidine kinase domains could reveal additional residues and mechanisms controlling TCS orthogonality.

3.2. Exploration of Non-Cognate Interactions

The ELIHKSIR web server allows for exploration and visualization of signaling networks. Using the displayed heatmap, users may identify crosstalk interactions in signaling networks. Non-cognate, crosstalk interactions are common in signaling networks and may influence the expression patterns in organisms. H_{TCS} scores can be used to identify non-cognate, crosstalk interactions. Non-cognate interactions may be predicted by high specificity for a non-cognate partner as shown in Figures 7b–d and 6b,d. Any negative score indicates some level of encoded specificity. While scores near zero indicate no encoded specificity, TCS non-cognate partners with scores near zero may still interact due to shared attributes present in all TCS proteins, shown in Figures 6c and 7b. TCS non-cognate pairs in which shared TCS attributes are partially removed have positive specificity scores indicating low specificity. These methods of identifying possible interactions may be used across all available organisms, allowing for users to investigate crosstalk interactions within specific, and possibly uncommon, species or strains.

TCS pairs in which the RR has a cognate HK of a different family than HisKA have low specificity, but may still interact are shown in Figures 7b,d,f and 6b,d,f. The ability to interact despite very low specificity indicates there may be activity of HATPase_c in phosphorylation of non-cognate RRs whose cognates belong to other HK families since HATPase_c is present in both HisKA and HisKA3 family HKs.

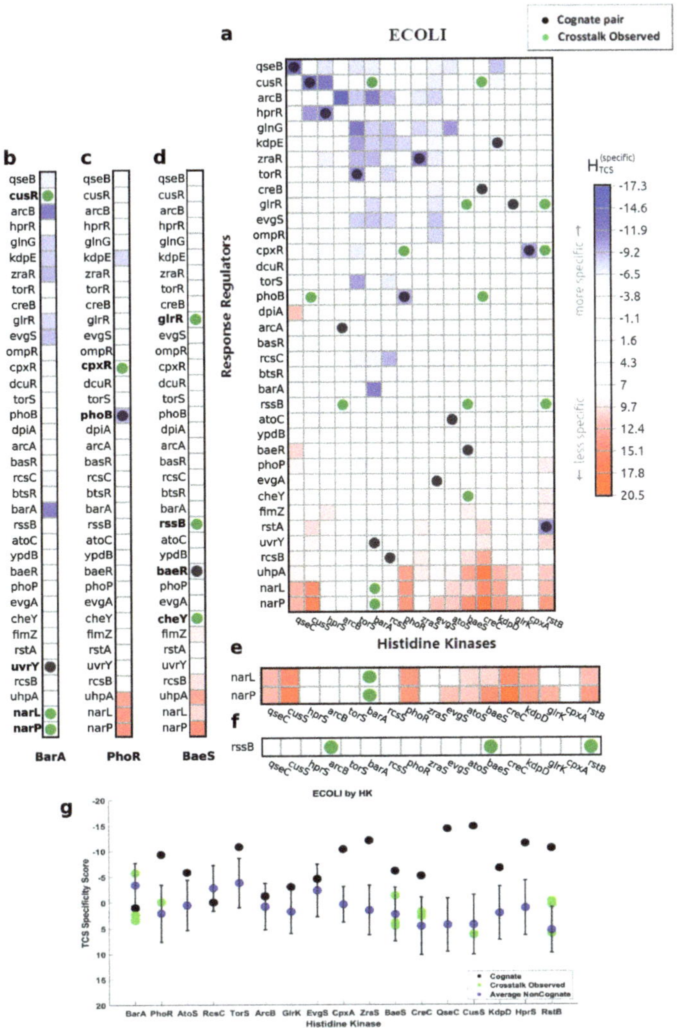

Figure 6. (**a**) Cognate interactions and observed in vitro crosstalk interactions overlaid onto the specificity score heatmap for *E. coli* [23]. Noncognate interactions are assessed. (**b**) BarA phosphorylates cusR, narL, and narP, in which the scores are −5.723, 2.390, and 3.491 respectively. The score for barA-cusR indicates that phosphorylation occurs due to high specificity for its noncognate partner. Phosphorylation of narL and narP are characterized in (**f**). (**c**) PhoR phosphorylates cpxR, in which the score is −0.037. A score near zero indicates diminished specificity, while still retaining attributes shared among all TCS pairs. (**d**) BaeS phosphorylates glrR, rssB, and cheY, in which the scores are −1.264, 3.998, and 4.605. The score for baeS-glrR indicates that phosphorylation occurs due to increased specificity for a noncognate partner. Phosphorylation of rssB is characterized in (**g**). Phosphorylation of cheY can be described similarly to (**f**), as its cognate HK utilizes a different family of HK than HisKA. (**e**) Cognate, crosstalk, and average non-cognate scores are shown for each HK. (**f**) HKs narQ and narX are not shown as they utilize a HisKA3 family HK, rather than HisKA. Their RRs, narL and narP, have low specificity for all HKs utilizing the HisKA domain. This leads narL and narP to be nonspecific for HisKA family HKs. Despite a lack of specificity, crosstalk is observed. (**g**) RssB is an orphan RR that can be phosphorylated by multiple HKs.

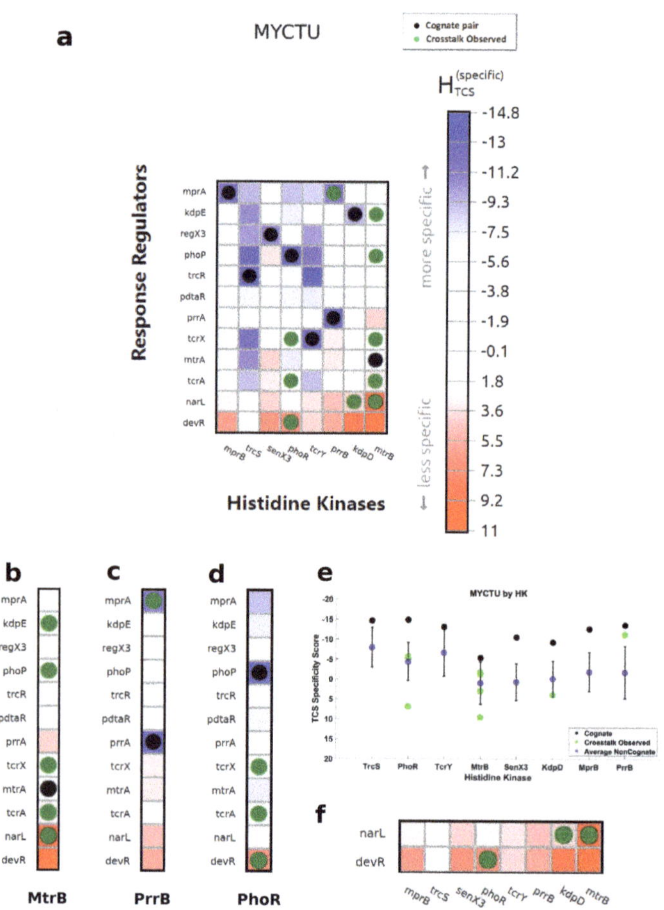

Figure 7. (**a**) Cognate interactions and observed in vitro crosstalk interactions overlaid onto th specificity score heatmap for M. tuberculosis [24]. Noncognate interactions are assessed. (**b**) Mtr phosphorylates kdpE, phoP, tcrX, tcrA, and narL, in which the scores are −4.895, −5.826, 0.391 −1.093, and 2.813 respectively. Scores for kdpE, phoP, and tcrA indicate that phosphorylatio by mtrB occurs due to high specificity for these noncognate partners. TcrX has a score near zero /textcolorredindicating diminished specificity but a presence of attributes shared among all TC pairs. Phosphorylation of narL is characterized in (**f**). (**c**) PrrB phosphorylates mprA, in which th score is −11.263. This score indicates that phosphorylation of mprA by prrB occurs due to high specificity. (**d**) PhoR phosphorylates tcrX, tcrA, and devR, in which the scores are −5.744, −5.17 and 6.856, respectively. Scores for tcrX and tcrA indicate that phosphorylation by phoR occurs du to high specificity for these noncognate partners. Phosphorylation of devR is characterized in (**f** (**e**) Cognate, crosstalk, and average noncognate scores are shown for each HK. (**f**) HKs devS and nar are not shown as they utilize a HisKA3 family HK, rather than HisKA. Their response regulator narL and devR, have low specificity for all HKs utilizing the HisKA domain.

In Figure 6g, we observe an orphan RR that exhibits low specificity for many HKs an has been phosphorylated by HKs with low predicted specificity. Aside from the possibilit of HATPase_c domain contributions, it is possible that low specificity for orphan RRs i favorable as it promotes promiscuity. In the case of rssB in *E. coli*, phosphorylation i important for function [25,26]. Therefore, promiscuity of rssB could ensure maintenance of function throughout the *E. coli* life cycle. Using similar reasoning, one can identify potential interactions with orphan HKs and RRs. Information yielded from analysis of orphan TCS proteins may assist in describing their role in organisms' life cycles, environmental stress responses, and expression patterns. Utilizing predicted orphan TCS protein interactions could be useful in the study of antibiotic resistance in bacteria, response to environmental metals and compounds in archaea, or plant response to drought.

3.3. Revealing Interaction Specificity for Mutation and Variation

After mutating a protein residue, specificity scores are recalculated and the heatmap is updated. This reveals how mutation(s) change interaction specificity with all possible TCS partners. A feature that becomes important when scientists would like to assess the network effect of mutations as opposed to single pairwise interactions. The ELIHKSIR web server also separates organisms by strain, allowing interaction specificities to be compared between different strains of the same organism. Accessibility of specificity predictions for different mutants and strains may reveal differences in TCS signaling of clinical and environmental variants and may assist in the engineering of sensory kinases and response regulators as it has been shown in previous studies [5].

4. Materials and Methods

4.1. MSA Construction

Raw HMM profiles for HisKA and REC were obtained through Pfam's hidden Markov models (HMM) [27,28]. Then, the profile was searched using Hmmer's hmmsearch against the TrEMBL database. HKs with a sequence gap of 5 residues or larger were excluded from the MSA. The resulting HisKA domain MSA was 67 residues in length and contained 111,032 sequences utilized in the ELIHKSIR web server. RRs with a sequence gap of 6 residues or larger were excluded from the MSA. The resulting REC domain MSA was 112 residues in length and contained 225,616 sequences utilized in the ELIHKSIR web server. Cognate HK-RR pairs were concatenated and used for the generation of couplings and local fields using mfDCA, where cognate is defined by having adjacent loci [29]. The resulting cognate MSA was 179 residues in length and contained 10,091 sequences. A number of 25 iterations of random concatenation of each HK to a random RR was used to generate a scrambled MSA. The resulting MSA was 179 residues in length and contained 16,363,100 sequences.

4.2. mfDCA Evolutionary Couplings and Hamiltonian Scores

Mean field DCA (mfDCA) [1] was used to infer the coevolutionary parameters from conjugated multiple sequence alignments (MSAs) of cognate HK–RR sequences and scrambled HK–RR sequences. The resulting coupling parameters and local field parameters were utilized in the calculation of Hamiltonian scores. In order to quantify changes on the Hamiltonian $H(S)$, Cheng et al. introduced a score H_{TCS} as follows:

$$H_{TCS}(HK_A + RR_A) = - \sum_{i=1}^{L_{HK_A}} \sum_{j=L_{HK_A}+1}^{L_{HK_A}+L_{RR_A}} e_{ij}(A_i, A_j) \times \Theta(c - r_{ij})$$
$$- \sum_{i=1}^{L_{HK_A}+L_{RR_A}} h_i(A_i)$$

$$(3)$$

for a specific pair between a sequence HK_A and RR_A of sequence lengths L_{HK_A} and L_{RR_A} with the coupling matrix $e_{ij}(A_i, A_j)$ between two sequence sites A_i, A_j at sequence positions i and j; and the local field $h_i(A_i)$ at the site A_i at sequence position i. L_{HK_A} is 67 for the HisKA domain and L_{RR_A} is 112 for the REC domain. The couplings are only taken within a pair distance $r_{ij} < c = 12\text{Å}$ of a native contact, expressed by a function $\Theta(x) = 1$ for all $x > 0$ and $\Theta(x) = 0$ for $x \leq 0$. The contact map of the native interfacial pairs is given by

the 3D resolved structure of protein interface Thermotoga maritima class I HK853 wi[th]
its cognate, RR468, (PDB ID: 3DGE). This interface is used as a template for the spati[al]
complex. Equation (3) is used to calculate energies H_{TCS} and H^0_{TCS} at interface positio[ns]
where H_{TCS} is calculated using cognate couplings and local fields and H^0_{TCS} is calculate[d]
using scrambled couplings and local fields. H^0_{TCS} is generated using the large-q Pot[ts]
Hamiltonian model on the scrambled MSA which is constructed by completing 25 roun[ds]
of concatenation of any of m HKs in the data set with any of n RRs in the dataset:

$$H^0_{\text{TCS}}(\{\text{HK}, \text{RR}\}) =$$
$$\langle H_{\text{TCS}}(\text{HK}_X | X \in \{1,...,m\} + \text{RR}_Y | Y \in \{1,...,n\}) \rangle_{25} \tag{(}$$

To find $H^{\text{specific}}_{\text{TCS}}$, Hamiltonian energies calculated from shared attributes present in a[ll]
HK–RR pairs must be removed from the specific HK–RR pair being evaluated:

$$H^{\text{specific}}_{\text{TCS}}(\text{HK}_A + \text{RR}_A) =$$
$$H_{\text{TCS}}(\text{HK}_A + \text{RR}_A) - H^0_{\text{TCS}}(\{\text{HK}, \text{RR}\}) \tag{(}$$

where the resulting $H^{\text{specific}}_{\text{TCS}}$ represents the interaction specificity strength between the H[K]
and RR. Therefore, this energy function could be used to predict the interaction preferen[ce]
between any HK and RR. Additionally, an updated $H^{\text{specific}}_{\text{TCS}}$ score, after incorporating
mutation in the MSA, serves a reference for the effect of the mutation on binding specifici[ty]
strength. The updated $H^{\text{specific}}_{\text{TCS}}$ is generated by performing the same calculations presente[d]
in Equations (3) and (5). Ranges for $H^{\text{specific}}_{\text{TCS}}$ values are varied between organisms an[d]
strains where a positive score indicates a loss of shared encoded TCS attributes, a negativ[e]
score indicates encoded specificity, and a score of zero indicates a presence of all share[d]
TCS attributes but diminished encoded specificity. When qualifying potential interactio[n]
users should compare $H^{\text{specific}}_{\text{TCS}}$ for different TCS pairs belonging to the same organis[m.]
One should consider more negative values to have increased encoded specificity, zero va[l-]
ues to be capable of interacting with other TCS proteins without encoded specificity in t[he]
HisKA domain, and positive values to exhibit insulation of HisKA and REC domains.

4.3. Software

The web server has a custom-built front end running React [30] for enhanced us[er]
experience with custom components. The back-end is serving data through REST [3[1]]
endpoints. Upon mutation, the scores are looked up from a pre-computed table. The pyth[on]
source code for the calculation of H_{TCS} is accessible via the web server. Details on publ[ic]
endpoints can be found in Appendix A.

5. Conclusions

The ELIHKSIR web server is a valuable tool for analyzing TCS specificity landscap[es]
in a growing list of 6412 species and strains of bacteria, 65 species and strains of archae[a,]
and 188 species and strains of eukaryotes. This allows users to find potential cross-ta[lk]
interactions and characterize existing orthogonality for many organisms across differe[nt]
kingdoms. For each organism, heatmaps and histograms of TCS networks are easily a[c-]
cessed, displayed, and exported. Furthermore, the ability to compute, display, and expo[rt]
changes in specificity for mutated HK or RR proteins allows users to explore potenti[al]
interactions and visualize changes in specificity over an entire signaling network. This ab[il-]
ity can assist in the analysis of engineered mutants, clinical and environmental variant[s,]
and cross-talk behavior. While ELIHKSIR is useful for interactions between HisKA fami[ly]
HKs and the REC domain of RRs, there exist other HK families in which the ELIHKS[IR]
model does not evaluate. Building and validating models to predict specificity for oth[er]
families of HK would further assist TCS research. Even though ELIHKSIR only displa[ys]
specificity scores for HisKA and REC domains, these domains are critical in determin[ing]

specificity for many TCS interactions, as demonstrated by the 6,272,607 HK-RR pairs evaluated. Due to the ability to mutate each protein and recalculate network-wide specificity scores, there are nearly endless possibilities of HK–RR pairs to evaluate using ELIHKSIR. The accessibility, breadth, and functionality of ELIHKSIR allows a variety of researchers (both computational and experimental) to harness TCS specificity predictions, supporting research efforts through a tool that did not previously exist.

Author Contributions: Conceptualization, F.M.; methodology, F.M., X.J. and C.S.; software, C.S. and Y.H.J.; validation, C.Z. and X.J.; formal analysis, C.Z., X.J., C.S.; investigation, C.Z., X.J., C.S.; resources, C.Z.; data curation, C.S. and C.Z.; writing—original draft preparation, C.S., C.Z.; writing—review and editing, F.M., C.Z., C.S.; visualization, C.S., C.Z.; supervision, F.M.; project supervision, F.M. All authors have read and agreed to the published version of the manuscript.

Funding: This research was funded by the University of Texas at Dallas (X.J. and F.M.); NIH grant number R35GM133631 (to C.S., X.J. and F.M.); and NSF grant number MCB-1943442 (to C.Z. and F.M.).

Institutional Review Board Statement: Not applicable.

Informed Consent Statement: Not applicable.

Data Availability Statement: Data presented in this work is available at the ELIHKSIR web server at https://elihksir.org/.

Conflicts of Interest: The authors declare no conflict of interest.

Abbreviations

The following abbreviations are used in this manuscript:

ELIHKSIR	Evolutionary Links Inferred for Histidine Kinase Sensors Interacting with Response regulators
TCS	Two-Component System
DCA	Direct-Coupling Analysis
mfDCA	couplings generated by mean-field method as outlined in Morcos, 2011 [1]
DI	Direct Information
HK	Histidine Kinase, Histidine Kinase family (Pfam:PF00512) [15]
RR	Response Regulator, Response Regulator family (Pfam: PF00072) [16]
TP	True Positive
FN	False Negative
PS	Positive Selection
NS	Negative Selection

Appendix A

Data of the server can be accessed in a programmatic way through two REST endpoints as described in Table A1. The all organisms endpoint `api/list` returns a list of all the organisms currently accessible through ELIHKSIR. The return value will contain the names (`ORGANISM_NAMES::STRING`), UNIPROT ID (`ORGANISM_UNIPROT_ID::STRING`), and the numeric identifier/primary key (`ORGANISM_ID::INT`) for each organism. By using the numeric identifiers obtained from the list endpoint further meta data and information, along with the scores for each interacting pair, can be obtained through the `api/pairs` endpoint.

Table A1. List of the available endpoints for the REST API.

Endpoint	HTTP Method	URL
All Organisms	GET	api/list
Pairs for heatmap	GET	api/pairs/{ORGANISM_ID::INT}

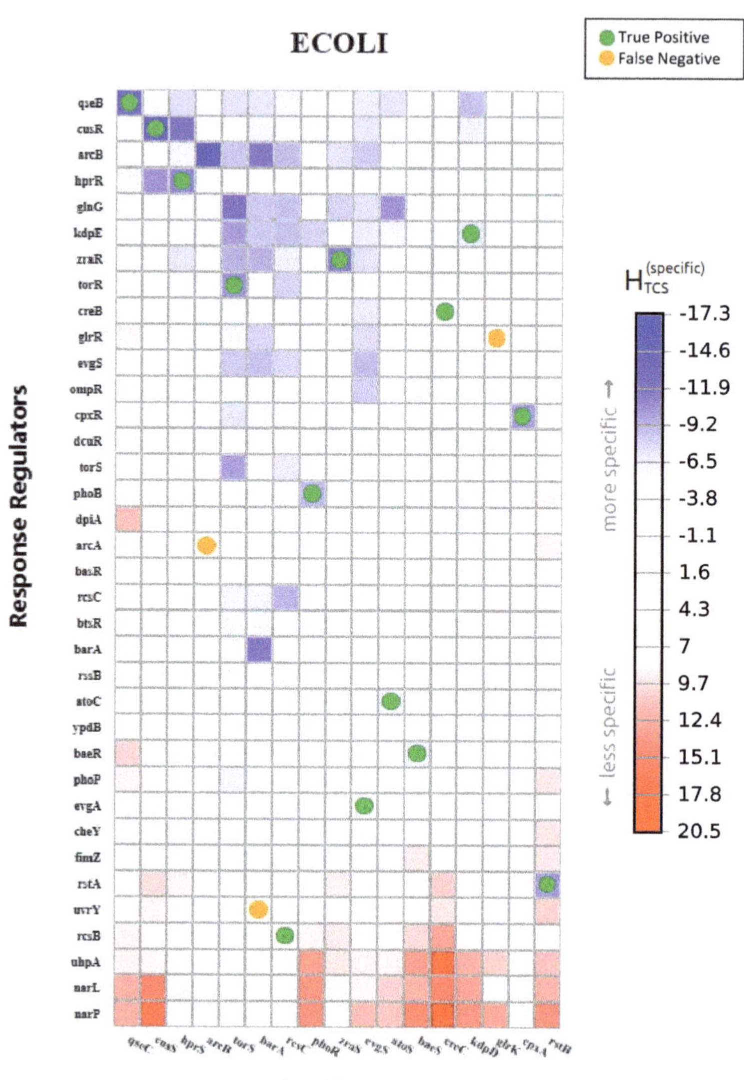

Figure A1. True positives are correct prediction of cognate pairs through positive and/or negativ selection. False negatives occur when the cognate pairing is not the most favorable interaction.

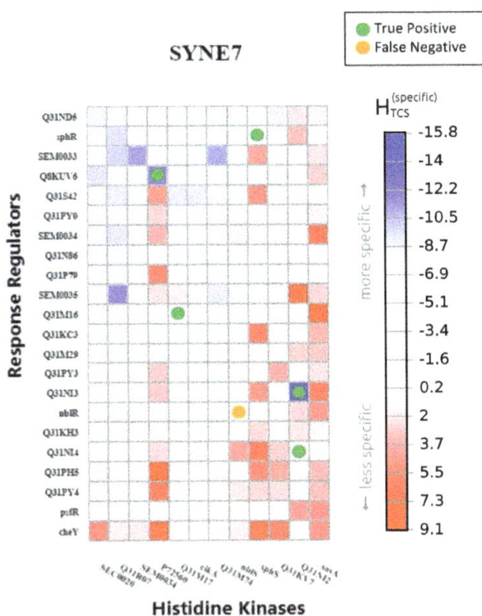

Figure A2. True positives are correct prediction of cognate pairs through positive and/or negative selection. False negatives occur when the cognate pairing is not the most favorable.

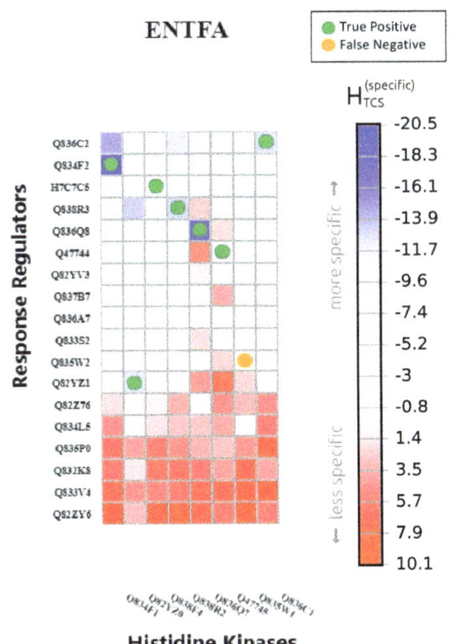

Figure A3. True positives are correct prediction of cognate pairs through positive and/or negative selection. False negatives occur when the cognate pairing is not the most favorable interaction.

Table A2. Couplings used in specificity model sorted by descending DI value.

HK	RR	DI	HK	RR	DI	HK	RR	DI	HK	RR	DI
18	77	0.102853	7	147	0.00806367	14	170	0.00572271	16	77	0.00460402
22	80	0.0722833	30	83	0.00801695	19	169	0.00571496	42	77	0.00460128
11	167	0.0705232	33	83	0.00781679	14	147	0.00570246	26	176	0.004455
26	84	0.0515243	22	171	0.00767719	22	170	0.00569941	42	76	0.00444185
23	80	0.0492594	19	79	0.00765574	16	76	0.00569788	17	169	0.00442802
14	146	0.04276	23	172	0.00763662	15	76	0.00568733	27	80	0.00440244
46	76	0.0398581	18	171	0.00759788	38	80	0.0056818	15	167	0.00439432
19	76	0.0392644	19	147	0.00753089	10	147	0.00567069	19	81	0.00420284
25	170	0.0351779	18	81	0.00752615	22	172	0.00562105	38	79	0.00418686
25	171	0.0303173	12	148	0.00734758	13	147	0.00559724	41	79	0.00416339
11	168	0.0285807	10	150	0.00732327	34	87	0.00559491	15	170	0.0041264
15	146	0.0270048	45	76	0.007169	33	84	0.00556154	24	173	0.00403743
29	87	0.0265711	22	76	0.00712089	34	84	0.00552466	18	146	0.00382329
19	77	0.0259669	21	172	0.00705718	25	169	0.00545037	16	147	0.00378105
30	87	0.0215653	30	80	0.00702085	15	118	0.00544512	18	172	0.00377972
23	76	0.0193616	15	74	0.00697282	25	174	0.00543429	20	168	0.00375797
19	80	0.0189693	18	147	0.00691656	18	74	0.00541696	16	169	0.00366294
22	77	0.0188355	45	79	0.00690392	14	149	0.00540391	25	81	0.00362641
23	79	0.0180874	22	78	0.00680398	30	86	0.00538702	13	169	0.00359782
19	74	0.0176283	19	170	0.00679327	33	87	0.00538074	10	148	0.00359015
8	147	0.0172729	23	170	0.00679065	17	147	0.00537632	20	77	0.00355346
18	169	0.0171606	18	78	0.00676363	33	86	0.00534217	11	146	0.00345648
29	171	0.0170736	26	81	0.00675062	7	149	0.00530426	21	81	0.00344168
16	168	0.0168674	31	87	0.0067342	14	169	0.00530394	42	80	0.00338203
15	77	0.0152404	21	77	0.00670382	38	83	0.00526882	28	173	0.00334077
25	172	0.0149784	27	84	0.00667465	26	82	0.00525117	22	174	0.0032904
39	83	0.014901	22	81	0.00666247	17	77	0.00524036	20	170	0.00327778
29	172	0.0146187	46	77	0.00658625	42	83	0.00521168	14	168	0.0032269
21	170	0.014469	26	79	0.00657677	34	83	0.00520958	22	176	0.00319442
26	80	0.0141692	18	168	0.00651457	34	80	0.00518935	24	172	0.00310293
26	83	0.0139579	45	80	0.00648245	20	80	0.00514101	17	170	0.00307046
23	83	0.0130196	19	172	0.00639985	46	80	0.00512939	19	168	0.00298315
12	168	0.0128269	18	76	0.00633046	18	170	0.00510377	26	85	0.00290492
15	147	0.0123301	28	172	0.00626959	23	82	0.00509604	20	171	0.0027667
29	84	0.0121863	25	175	0.00623035	25	80	0.00502442	15	169	0.00275228
8	148	0.0119021	16	74	0.00615123	49	77	0.00502117	20	169	0.00269252
23	84	0.0118386	30	88	0.00614559	45	77	0.00501361	15	168	0.00266174
22	84	0.0113964	19	75	0.00610985	24	171	0.00496918	21	168	0.00254422
23	171	0.0108972	10	149	0.00608531	17	76	0.00494564	26	174	0.00238278
32	87	0.0108637	24	80	0.00607124	14	167	0.00492586	27	173	0.00238187
26	171	0.0107505	23	169	0.00605014	15	148	0.00492344	24	169	0.00237797
25	84	0.0104978	33	88	0.00604033	23	173	0.00489615	20	172	0.00233119
14	148	0.0104055	7	150	0.00599047	24	170	0.00488941	18	145	0.00222389
30	84	0.0102802	48	76	0.00596518	44	76	0.00484498	13	168	0.00190674
26	87	0.0102777	21	80	0.0059512	21	173	0.00483877	10	169	0.00186603
23	78	0.01007	25	173	0.00592837	29	173	0.00481891	18	167	0.00162171
23	77	0.00997545	28	171	0.00591566	10	168	0.00480323	12	169	0.00147153
22	169	0.00996165	42	79	0.00586154	17	171	0.00478502	15	145	0.00135628
43	80	0.00972282	22	173	0.00585409	22	168	0.00478308	11	149	0.00105631
22	83	0.00952424	26	172	0.0058532	12	147	0.0047804	11	169	0.000969333
18	80	0.0093605	49	76	0.005838	26	173	0.00476983	11	150	0.000862052

Table A2. *Cont.*

HK	RR	DI	HK	RR	DI	HK	RR	DI	HK	RR	DI
7	148	0.00897782	26	175	0.00581626	22	82	0.00476307	11	148	0.000856593
29	83	0.00882006	30	172	0.00577961	45	75	0.00475623	11	118	0.000790599
21	171	0.00865425	22	79	0.00577265	39	80	0.00473036	11	147	0.000413171
26	170	0.00854487	15	73	0.00576882	27	172	0.00472585			
19	171	0.00831504	41	80	0.00576018	41	76	0.00472158			
17	168	0.008266	30	173	0.00574398	27	83	0.0046522			
19	78	0.00818936	23	81	0.00573789	20	76	0.00465154			
21	169	0.0080992	22	175	0.00573107	16	77	0.00460402			

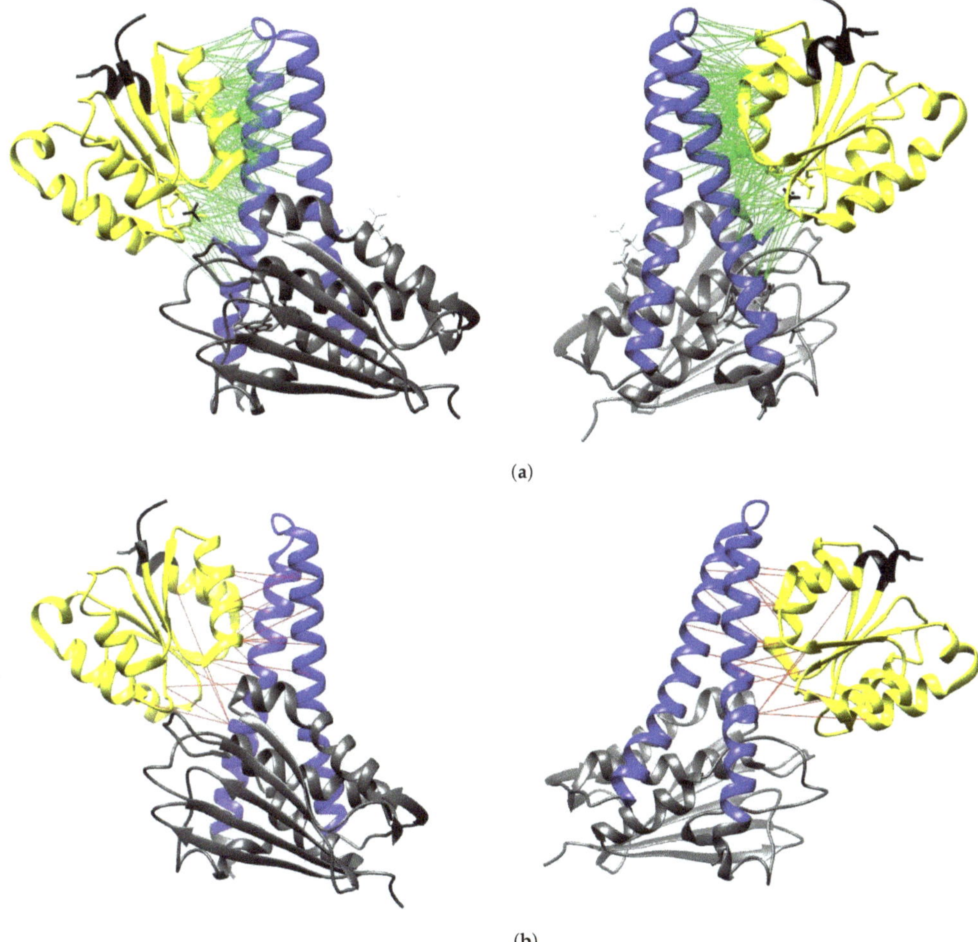

(a)

(b)

Figure A4. Gray structures show the HK residues lying outside of the HisKA domain. Black structures show the RR residues lying outside the REC domain. The blue structure represents the HisKA domain, and the yellow structure represents the REC domain. Green pseudobonds show contacts within 12 Angstroms C_α to C_α. Red pseudobonds show the top 20 DCA couplings. The distribution of DCA couplings indicates that the model does not show biases towards subregions of the interface. (**a**) All contacts within 12 Angstroms as found in the structure viewed from two different positions, left and right faces; (**b**) Top 20 interfacial DI contacts as viewed from left and right faces.

References

1. Morcos, F.; Pagnani, A.; Lunt, B.; Bertolino, A.; Marks, D.S.; Sander, C.; Zecchina, R.; Onuchic, J.N.; Hwa, T.; Weigt, M. Direct coupling analysis of residue coevolution captures native contacts across many protein families. *Proc. Natl. Acad. Sci. USA* **201** *108*. [CrossRef] [PubMed]
2. Cheng, R.R.; Morcos, F.; Levine, H.; Onuchic, J.N. Toward rationally redesigning bacterial two-component signaling system using coevolutionary information. *Proc. Natl. Acad. Sci. USA* **2014**, *111*. [CrossRef] [PubMed]
3. Boyd, J.S.; Cheng, R.R.; Paddock, M.L.; Sancar, C.; Morcos, F.; Golden, S.S. A combined computational and genetic approach uncovers network interactions of the cyanobacterial circadian clock. *J. Bacteriol.* **2016**, *198*, 2439–2447. [CrossRef] [PubMed]
4. Cheng, R.R.; Nordesjö, O.; Hayes, R.L.; Levine, H.; Flores, S.C.; Onuchic, J.N.; Morcos, F. Connecting the sequence-space of bacterial signaling proteins to phenotypes using coevolutionary landscapes. *Mol. Biol. Evol.* **2016**, *33*, 3054–3064. [CrossRef] [PubMed]
5. Cheng, R.R.; Haglund, E.; Tiee, N.S.; Morcos, F.; Levine, H.; Adams, J.A.; Jennings, P.A.; Onuchic, J.N. Designing bacterial signaling interactions with coevolutionary landscapes. *PLoS ONE* **2018**, *13*, e0201734. [CrossRef]
6. Morcos, F.; Hwa, T.; Onuchic, J.N.; Weigt, M. Direct Coupling Analysis for Protein Contact Prediction. In *Protein Structure Prediction*; Springer: New York, NY, USA, 2014; pp. 55–70. [CrossRef]
7. Muscat, M.; Croce, G.; Sarti, E.; Weigt, M. FilterDCA: Interpretable supervised contact prediction using inter-domain coevolution. *bioRxiv* **2019**. Available online: https://www.biorxiv.org/content/early/2019/12/24/2019.12.24.887877.full.pdf (accessed on 11 December 2020).
8. Szurmant, H.; Weigt, M. Inter-residue, inter-protein and inter-family coevolution: bridging the scales. *Curr. Opin. Struct. Biol.* **2018**, *50*, 26–32. [CrossRef]
9. Jiang, X.L.; Martinez-Ledesma, E.; Morcos, F. Revealing protein networks and gene-drug connectivity in cancer from direct information. *Sci. Rep.* **2017**, *7*, 3739. [CrossRef]
10. Jacquin, H.; Gilson, A.; Shakhnovich, E.; Cocco, S.; Monasson, R. Benchmarking Inverse Statistical Approaches for Protein Structure and Design with Exactly Solvable Models. *PLoS Comput. Biol.* **2016**, *12*, e1004889. [CrossRef]
11. Levy, R.M.; Haldane, A.; Flynn, W.F. Potts Hamiltonian models of protein co-variation, free energy landscapes, and evolutionary fitness. *Curr. Opin. Struct. Biol.* **2017**, *43*, 55–62. [CrossRef]
12. Figliuzzi, M.; Jacquier, H.; Schug, A.; Tenaillon, O.; Weigt, M. Coevolutionary landscape inference and the context-dependence of mutations in beta-lactamase tem-1. *Mol. Biol. Evol.* **2016**, *33*, 268–280. [CrossRef] [PubMed]
13. Ekeberg, M.; Hartonen, T.; Aurell, E. Fast pseudolikelihood maximization for direct-coupling analysis of protein structure from many homologous. *J. Comput. Phys.* **2014**, *276*, 341–356. [CrossRef]
14. El-Gebali, S.; Mistry, J.; Bateman, A.; Eddy, S.R.; Luciani, A.; Potter, S.C.; Qureshi, M.; Richardson, L.J.; Salazar, G.A.; Smart, A.; et al. The Pfam protein families database in 2019. *Nucleic Acids Res.* **2018**, *47*, D427–D432. Available online: https://academic.oup.com/nar/article-pdf/47/D1/D427/27436497/gky995.pdf (accessed on 11 December 2020). [CrossRef] [PubMed]
15. Pfam. *Family: HisKA (PF00512)—His Kinase A (Phospho-Acceptor) Domain*; Pfam: Hinxton, UK, 2020.
16. Pfam. *Family: Response_reg (PF00072) Response Regulator Receiver Domain*; Pfam: Hinxton, UK, 2020.
17. Consortium, T.U. UniProt: A worldwide hub of protein knowledge. *Nucleic Acids Res.* **2018**, *47*, D506–D515. [CrossRef]
18. Schug, A.; Weigt, M.; Onuchic, J.N.; Hwa, T.; Szurmant, H. High-resolution protein complexes from integrating genomic information with molecular simulation. *Proc. Natl. Acad. Sci. USA* **2009**, *106*, 22124–22129. [CrossRef]
19. Szurmant, H.; Hoch, J.A. Interaction fidelity in two-component signaling. *Curr. Opin. Microbiol.* **2010**, *13*, 190–197. [CrossRef]
20. Capra, E.J.; Laub, M.T. The Evolution of Two-Component. *Annu. Rev. Microbiol.* **2012**, *66*, 325–347. [CrossRef]
21. Heath, J.D.; Charles, T.C.; Nester, E.W. Ti Plasmid and Chromosomally Encoded Two-Component Systems Important in Plant Cell Transformation by Agrobacterium Species. In *Two—Component Signal Transduction*; John Wiley & Sons, Ltd.: Hoboken, NJ, USA, 1995; Chapter 23, pp. 367–385.
22. Stewart, R.C.; Jahreis, K.; Parkinson, J.S. Rapid phosphotransfer to CheY from a CheA protein lacking the CheY-binding domain. *Biochemistry* **2000**, *39*, 13157–13165. [CrossRef]
23. Yamamoto, K.; Hirao, K.; Oshima, T.; Aiba, H.; Utsumi, R.; Ishihama, A. Functional characterization in vitro of all two-component signal transduction systems from Escherichia coli. *J. Biol. Chem.* **2005**, *280*, 1448–1456. [CrossRef]
24. Agrawal, R.; Pandey, A.; Rajankar, M.P.; Dixit, N.M.; Saini, D.K. The two-component signalling networks of Mycobacterium tuberculosis display extensive cross-talk in vitro. *Biochem. J.* **2015**, *469*, 121–134. [CrossRef]
25. Becker, G.; Klauck, E.; Hengge-Aronis, R. Regulation of RpoS proteolysis in Escherichia coli: The response regulator RssB is recognition factor that interacts with the turnover element in RpoS. *Proc. Natl. Acad. Sci. USA* **1999**, *96*, 6439–6444. [CrossRef] [PubMed]
26. Klauck, E.; Lingnau, M.; Hengge-Aronis, R. Role of the response regulator RssB in σS recognition and initiation of σS proteolysis in Escherichia coli. *Mol. Microbiol.* **2001**, *40*, 1381–1390. [CrossRef] [PubMed]
27. Eddy, S.R. Profile hidden Markov models. *Bioinformatics* **1998**, *14*, 755–763. [CrossRef] [PubMed]
28. Finn, R.D.; Mistry, J.; Tate, J.; Coggill, P.; Heger, A.; Pollington, J.E.; Gavin, O.L.; Gunasekaran, P.; Ceric, G.; Forslund, K.; et al. The Pfam protein families database. *Nucleic Acids Res.* **2009**, *38*, 211–222. [CrossRef] [PubMed]
29. Williams, R.H.; Whitworth, D.E. The genetic organisation of prokaryotic two-component system signalling pathways. *BMC Genom.* **2010**, *11*, 720. [CrossRef]

Facebook Inc. *React—A JavaScript Library for Building User Interfaces*; Facebook Inc.: Menlo Park, CA, USA, 2020.
Fielding, R.T.; Taylor, R.N. Architectural Styles and the Design of Network-Based Software Architectures. Ph.D. Thesis, University of California, Irvine, CA, USA, 2000.

MDPI
St. Alban-Anlage 66
4052 Basel
Switzerland
Tel. +41 61 683 77 34
Fax +41 61 302 89 18
www.mdpi.com

Entropy Editorial Office
E-mail: entropy@mdpi.com
www.mdpi.com/journal/entropy